Dispatches from Continent Seven

Dispatches
from
Continent
Seven

AN ANTHOLOGY OF ANTARCTIC SCIENCE

REBECCA PRIESTLEY

AWA PRESS

First edition published in 2016 by Awa Press, Unit 1, Level 3,
11 Vivian Street, Wellington 6011, New Zealand.

ISBN 978-1-927249-05-5

Ebook formats
Epub 978-1-927249-42-0
Mobi 978-1-927249-43-7

Cover photograph by NASA Earthdata

Cover design by Greg Simpson
Typesetting by Tina Delceg
This book is typeset in Sabon, Baskerville, Vista Slab and National

Printed by 1010 Printing Group Ltd, China

Find more great books at awapress.com.

Produced with the assistance of

Antarctica New Zealand

For Peter Barrett,
the first Antarctican I ever met

Contents

x Contributors

xiv The frozen pages *Gregory O'Brien*

xxi Introduction

UNKNOWN LAND

1 Cook circumnavigates the continent *James Cook*

13 Pickled penguins and barrels of ice
Thaddeus von Bellingshausen

21 Drinking wine in Adélie Land *Joseph Dubouzet*

27 Wilkes among the ice islands *Charles Wilkes*

37 Fire and ice *James Clark Ross*

43 The polar captain's wife *Chris Orsman*

THE FIRST ANTARCTICANS

47 In a sleeping bag beneath the aurora australis
Frederick Cook

53 The art and science of sledge travel *Leopold McClintock*

63 Skinned penguins and bloody seals *Edward Wilson*

79 Drygalski's balloon ascent *Erich von Drygalski*

85 Luxuriant vegetation and extensive coasts
Otto Nordenskjöld

91 Penguin eggs and fried seal *Johan Gunnar Andersson*

97 Hunting the south magnetic pole *Edgeworth David*

111 Nematodes, rotifers, water bears and mites *James Murray*

117 The measure *Alice Miller*

119 The worst journey in the world *Apsley Cherry-Garrard*

135   The hooligan cocks of Cape Adare   *George Murray Levick*

141   George Murray Levick and the Adélie penguins   *Helen Heath*

143   Impressions   *Robert Falcon Scott*

147   Geologising on the Beardmore   *Robert Falcon Scott*

159   Ice monsters, growlers and bergy bits   *Raymond Priestley*

175   Byrd makes a meteorological observation   *Richard Byrd*

187   Crabeaters and leopards   *Graham Turbott*

CONTINENT FOR SCIENCE

197   Innocents in the Dry Valleys   *Colin Bull*

217   Birds and mammals of Antarctica   *Bernard Stonehouse*

223   Catching falling stars   *William A. Cassidy*

231   The accidental penguin biologist   *Lloyd Spencer Davis*

236   Food chain   *Bill Manhire*

239   Krill   *David G. Campbell*

248   Small fry   *Ashleigh Young*

251   An average day in the deep field   *John Long*

261   Water, ice and stone   *Bill Green*

269   The lakes of Mars   *Chris Orsman*

273   Katabatic winds   *Stephanie Shipp*

277   Inside the emperor penguin egg   *Gavin Francis*

281   The mountains under the ice   *Robin Bell*

289   Antarctic time capsule   *Michael S. Becker*

293   Neutrinos on ice   *Katie Mulrey*

GLOBAL BAROMETER

307   Earth sans sunscreen   *Jonathan Shanklin*

313   Waiting for the polar sunrise   *Rhian Salmon*

329   Life under ice   *Paul Dayton and Simon Thrush*

335 Thirty-six million years in Antarctica   *Rob Dunbar*

347 Hermaphrodite butterflies and acid seas   *James McClintock*

357 Fishing in Antarctica   *Victoria Metcalf*

367 Sea ice and polynyas   *Helen Bostock*

375 A page from the ice diary   *Nancy Bertler*

381 What do seals want for Christmas?   *Regina Eisert*

389 March of the king crabs   *Kathryn Smith*

395   Glossary of scientific and nautical terms

401   Sources and copyright

406   Illustration credits

411   Acknowledgements

413   Index

Contributors

Andersson, Johan Gunnar (1874–1960) Swedish geologist who was a member of the Swedish Antarctic Expedition of 1901–03.

Becker, Michael S. (1986–) US ecologist, scientific diver and photographer specialising in polar environments.

Bell, Robin (1958–) US marine geophysicist who has coordinated many expeditions to Antarctica to study ice-sheet dynamics and sub-glacial topography.

Bellingshausen, Thaddeus von (1778–1852) German officer in the Imperial Russian Navy whose 1919–21 expedition was one of the first to sight the Antarctic continent.

Bertler, Nancy (1970–) German-born glaciologist, now leading ice core investigations in Antarctica to reveal the history of Antarctic climate over the last 100,000 years.

Bostock, Helen (1977–) Australian-born marine geologist, now working in New Zealand, with research interest in analysing sediment from the sea floor to understand past changes in the Southern Ocean.

Bull, Colin (1928–2010) UK-born physicist who led a two-month Victoria University of Wellington expedition to the McMurdo Dry Valleys in 1958–59, returned to Antarctica with the university's 1960–61 Antarctic expedition, and later helped establish the Institute of Polar Studies at Ohio State University.

Byrd, Richard Evelyn (1888–1957) US naval officer, explorer and aviator who made five trips to Antarctica, on the second of which, in 1933–35, he spent a winter alone operating a remote meteorological station.

Campbell, David G. (1949–) US ecologist who has conducted research into krill and other marine invertebrates in Antarctic waters.

Cassidy, William A. US meteoriticist who founded the US Antarctic Search for Meteorites Project (ANSMET).

Cherry-Garrard, Apsley (1886–1959) Untrained assistant zoologist on Robert Falcon Scott's second Antarctic expedition, 1910–13.

Cook, James (1728–1779) British explorer and captain in Royal Navy who undertook three voyages of exploration to the Pacific and was the first to circumnavigate Antarctica.

Cook, Frederick (1865–1940) US explorer, physician and ethnographer on Adrien De Gerlache's Belgian Antarctic Expedition, 1897–99.

David, Edgeworth (1858–1934) Welsh-born Australian geologist and member of Ernest Shackleton's 1907–09 Antarctic expedition who led the first ascent of Mount Erebus and the first trek to the south magnetic pole.

Davis, Lloyd Spencer (1954–) New Zealand zoologist and science communicator who specialises in the study of birds and mammals, especially penguins.

Dayton, Paul (1941–) US marine ecologist and biological oceanographer who began working in Antarctica in 1963, focusing on research into coastal habitats, and has carried out more than 500 dives under the ice in McMurdo Sound.

Drygalski, Erich von (1865–1949) German geographer and geophysicist who led the first German expedition to Antarctica, 1901–03.

Dumont d'Urville, Jules (1790–1842) French explorer and naval officer who led a French expedition around the world and to Antarctica in 1837–40.

Dunbar, Rob (1954–) US earth systems scientist whose scientific interests include climate dynamics, oceanography, marine ecology and biogeochemistry.

Eisert, Regina (1972–) German marine biologist with a particular research interest in Antarctica megafauna, including whales, seals and toothfish.

Francis, Gavin (1975–) Scottish physician who, while stationed during 2003–04 at Halley Research Station, the British Antarctic Survey base in East Antarctica, carried out research on emperor penguins.

Green, Bill (1942–) US limnologist and geochemist who began researching landlocked lakes in the McMurdo Dry Valleys of Antarctica in the 1980s.

Levick, George Murray (1876–1956) Royal Navy surgeon and zoologist on Robert Falcon Scott's 1910–13 Antarctic expedition.

Long, John (1957–) Australian palaeontologist and science communicator specialising in fossil fish, including those found in Antarctica.

Contributors

McClintock, James (1955–) US biologist specialising in Antarctic marine biology and the effects of climate change on Antarctica.

McClintock, Leopold (1819–1907) Irish-born British naval officer and Arctic explorer who pioneered long-distance polar sledge travelling involving man-hauling, thus also assisting Antarctic explorers.

Metcalf, Victoria (1974–) New Zealand marine biologist and science communicator focusing on Antarctic fish.

Mulrey, Katie (1985–) US astrophysics postgraduate student studying cosmic rays and neutrinos with ANITA project in Antarctica, 2014–15.

Murray, James (1865–1914) Scottish biologist specialising in tardigrades; member of Ernest Shackleton's 1907–09 Antarctic expedition.

Nordenskjöld, Otto (1869–1928) Geologist who led a Swedish scientific expedition to the Antarctic Peninsula from 1901–04.

Priestley, Raymond (1886–1974) British geologist who was a member of both Ernest Shackleton's 1907–09 Antarctic expedition and Robert Falcon Scott's second Antarctic expedition, 1910–13.

Ross, James Clark (1800–1862) British naval officer and explorer who led a voyage of scientific discovery to Antarctica and the Southern Ocean from 1839 to 1843.

Salmon, Rhian (1974–) UK atmospheric chemist who spent three summers and a winter at Halley Research Station commissioning a "clean air" laboratory and implementing a project to measure trace gases in the troposphere.

Scott, Robert Falcon (1868–1912) British naval officer and explorer who led two expeditions to Antarctica, 1901–04 and 1910–13.

Shanklin, Jonathan (1953–) Meteorologist with the British Antarctic Survey since 1977 and one of the three-person team who discovered the ozone hole above Antarctica in the 1980s.

Shipp, Stephanie US glacial geologist who has conducted research into how Antarctic ice sheets have changed over the last 20,000 years.

Smith, Kathryn (1980–) UK marine and evolutionary biologist whose interests include the effects of invading predators on Antarctic creatures living on or in the sea floor due to climate change.

Stonehouse, Bernard (1926–2014) UK biologist who studied birds and mammals in Antarctica, including an eighteen-month study of emperor penguins on South Georgia, and researched the impact of human activity on Antarctic plants, animals and soils.

Thrush, Simon (1958–) New Zealand marine ecologist specialising in coastal ecosystems, including those in Antarctica.

Turbott, Graham (1914–2014) New Zealand zoologist, ornithologist and entomologist who undertook field trips to subantarctic islands.

Wilkes, Charles (1798–1877) US naval captain and leader of the United States Exploring Expedition of 1838–42.

Wilson, Edward (1872–1912) British zoologist, surgeon and artist on the 1901–04 and 1911–13 Antarctic expeditions led by Robert Falcon Scott.

POETS

Heath, Helen (1970–) Poet whose poem in this volume was inspired by reading George Murray Levick's account of *The Sexual Habits of the Adélie Penguin*.

Manhire, Bill (1946–) Poet, short story writer, founding director of International Institute of Modern Letters at Victoria University of Wellington, and editor of *The Wide White Page: Writers Imagine Antarctica*; travelled to Antarctica in 1998 on Antarctica New Zealand's inaugural Artists to Antarctica programme.

Miller, Alice (1982–) Poet who visited Antarctica in summer of 2011, staying at Scott Base and visiting Scott's and Shackleton's huts and a US field base in the Taylor Valley.

O'Brien, Gregory (1961–) Poet, writer, artist and essayist.

Orsman, Chris (1955–) Poet who travelled to Antarctica in 1998 on Antarctica New Zealand's inaugural Artists to Antarctica programme.

Young, Ashleigh (1983–) Poet, essayist and editor, whose poem in this volume was inspired by reading David G. Campbell's "Krill" from his book *The Crystal Desert: Summers in Antarctica*.

The frozen pages

For Rebecca Priestley

How is it, Rebecca, the ice shelf
becomes eventually a shelf
    of books; and what do we make,

        from this distance,
    of that unimaginable
whiteness? Is there a science

of not knowing
    as there is a poetry of
        unimagining? The closest I ever came

    to your frozen continent was
the other end of
the whale highway –

    a hotel foyer in Nuku'alofa
        to be precise –
    arriving by sea upon

that offshore ice-box *Otago*,
fresh from the Subantarctic,
    with its occasional crew

        of poets, painters, a solar panel-man
    and government minister.
Departing northwards from Devonport

a carton of books left
    on the helicopter deck
        – our offering to the Tongan people –

was quickly spirited
below deck. By sea, as you know, a direct passage
can be a twisted, convoluted thing,

    a process of endless realignment – no one
      expects anything to be
    where it was. And so, upon docking
in Nuku'alofa,

the carton was nowhere
    to be found, the vessel turned subsequently
      upside down, the fury

    of the Supplies Officer and the evacuation
of District Nine
to no avail.

    A further week on, however,
      a naval rating came knocking
    at our hotel, placed a dark, rectangular

object just inside the office door – and this
is where Antarctica enters
    our poem – our missing box and

    its contents. Frozen solid.
    Departing Auckland harbour,
the sailor explained to us

he had been assigned the package, but
    not knowing what it contained
      and worried the contents

    might be perishable, he stowed it
in the deep freezer – a reconditioned
shipping container on the aft deck –

where it vanished
   behind a wall of frozen meat,
beneath a layer of powdery ice.

Harbourside in Nuku'alofa,
one of the ship's cooks
   stumbled upon it. And now,

      under its small cloud of steam,
   the box was left to thaw
in the tropical heat, in the company

of luxuriant palms, lizards and,
   by evening, fruit bats. An iceberg
      adrift
   on the overnight floor

of Waterfront Lodge − while
we pondered our ruined consignment, how
   ice crystals might merge

      with the written word and, later, how
            we might disentangle
   these hybrid structures. The following morning
revealed, happily, only

a slight buckling, the pages
   like puckered skin,
      the books revived, their ice journey

   now far behind them. And there, Rebecca,
you have it: our frozen library
a replica or emblem of your

   beloved continent, with its
      accumulated intelligence of water, ice
   and air. I think it was

Kafka who wrote 'a book
should serve as an axe
   for the frozen sea

     within us'. Only here the books
  have become
the frozen sea, which means,

in all probability, humanity must be
   the axe – which, as you know, can sometimes seem
     alarmingly the case. There was one other

   polar apparition, this time
flying southwards
to Dunedin, November 1996,
  shortly after two hundred

    icebergs came famously unstuck
  from the Ronne Shelf.
Our pilot detoured seawards

so we could take in the kilometre-long chunk
   of whiteness – the first iceberg within sight
    of mainland New Zealand,

  we were told, since 1931 – its
surface as flat as a book cover, its heft that
of a well-stocked encyclopaedia.

   Within days, a film crew from Sydney
    had made a raucous landing
  by helicopter

and planted an Australian flag; later
some locals offloaded and set to shearing a sheep.
  What can I say? I'd rather dwell on the fact

that ice has something to
tell us – just as it spoke to the young painter Hanly
at a Kāpiti Coast health camp

in the 1930s, when he spied
a translucent triangle
on the horizon. And

from that vision (no official sighting
was recorded) his career as a painter
sprung. I have in mind

your southern continent with
its creaking and popping,
its groaning and inexhaustible

quiet – those great dry valleys of its
unvisiting – but it is these far satellites
that reach me:

Hanly's precious iceberg – like
crystal, like Waterford –
and the frozen bookshelf of HMNZS *Otago*.

In face of the prevailing
uncertainties and a family devoted to
more conventional farming, a cousin of mine

once divided his time – six months to be
precise – between water and ice, installing
a hydroponics system at Scott Base –

but that season is remembered mainly for
the morning his tractor fell through pack ice,
plummeting twenty metres

to the seabed, from where he managed
to unbuckle himself and somehow find
the opening above.

Betwixt ice-box and berg, daily melt and
cracked, frozen sea, we surface
each morning; we come around

as by Antarctic Circumpolar Current
    to whatever fresh predicament
        might befall

    this planet of our undreaming
and, hopefully, not our undoing – such
matters as tumble or fall

    from book or polar
        shelf. Such things as are ours
    not to know, Rebecca,

our planet
out on its furthest, frozen limb –
    to see our crystalline faces in it, steady

        ourselves upon it and
    in these days of thinning
polar ice, not fall through.

Gregory O'Brien

Introduction

They say for every 10°C drop in temperature it's "a different kind of cold". At minus 20°, on a 1,300-metre-high, ice-free plateau in the Dry Valleys region of the Transantarctic Mountains, I have hit my limit. I ache with the cold. I am walking and talking more slowly. I am constantly out of breath. My job – operating a camera while my colleague Cliff Atkins gives a series of filmed lectures on Antarctic geology and paleoclimate – is passive: once the camera is rolling I just have to stand still and keep quiet. And when the wind blows from the south I feel even colder.

The geologists I am with at Friis Hills seem unconcerned by the conditions. They are veterans of Antarctic science – among the nine of us we have seen fifty-two Antarctic seasons – and their minds are elsewhere. Here to investigate times in Earth's history when the continent was significantly warmer than it is today, they are focused on the sediments and glacial deposits beneath our boots.

While they look down, I look up and out. From our campsite – a scattering of yellow and orange sleeping tents, and a blue Polar Haven mess tent – I can see south over the Taylor Glacier towards the Kukri Hills, whose steep cliffs show stripes of granite from 500 million years ago and dolerite sills from 180 million years ago. To the northwest, through a gap in the rocks on the near horizon, is a glimpse of the endless white of the Polar Plateau.

Adam Lewis, an American glacial geomorphologist, is the member of our team most familiar with this environment – he has spent five previous field seasons here. To me, the hills are a landscape of brown on brown: hip-high, wind-blasted boulders looming over a ventifact pavement of tightly packed flat-surfaced rocks in a matrix of sand. To Adam, they are clues to millions of years of history.

Rebecca Priestley

On his first visit, he and a colleague found fossil tundra everywhere they looked. The leaves, insects, pollen and seeds from the sediments below the rocks hinted at a climate with summer air temperatures some 20°C warmer than today. He and his students have since identified a sequence of glacial moraines that were once interspersed with life-supporting water bodies – ponds, marshes and small lakes – created by glacial meltwaters. It's clear that the Friis Hills landscape was created by a series of advancing and retreating glaciers, but one of the things that's remarkable about it, and makes it so interesting to geologists, is that it was created twenty to fourteen million years ago and has remained intact since then. Evidence suggests that about fourteen million years ago the climate changed: air temperatures became colder, rain stopped falling, and the water became locked up in glacial ice. Plants could no longer grow here.

Even one day at Friis Hills gives me a new-found respect for Antarctica's early visitors, the heroic age explorers, and the scientists who followed them. After our first day's work geologising and filming we meet in the Polar Haven, crammed around a small table and vying for position nearest the stove that cooks our food and warms the air. But no matter how warm it gets inside – the temperature reaches 7°C – the cold rises up through my boots, into my feet and up my legs.

As we shelter in the tent it begins to snow and the brown Friis Hills landscape starts turning white. I know that even if there were something seriously wrong with me – *Why can't I breathe?* – I would be stuck here. Initially it had been reassuring to know we were only a forty-minute helicopter ride from Scott Base, but I now realise that in this Condition One weather, with blowing snow and almost no visibility, no one can come and get me. The fact there is nowhere to go – *I've had enough now, let me inside please* – gives me a sense of claustrophobia and an anxiety I work hard to keep in check.

In 1912, nearly a hundred years before my first visit to Antarctica, Robert Falcon Scott and the other four men in his polar party – Henry Bowers, Edgar Evans, Lawrence Oates and Edward Wilson – marched, camped and died on the Ross Ice Shelf in conditions much worse than those I am experiencing. The place where they died, some 250 kilometres south of Ross Island, is now covered in decades of accumulated snow and ice, and slowly moving towards the edge of the ice shelf and the open sea.

A century on, travelling to and surviving in this continent has become a very different experience. It took Scott months to sail from the UK to Antarctica, and on the final leg of the journey, between New Zealand and Cape Evans, his reinforced wooden ex-whaling ship *Terra Nova* endured a fierce Southern Ocean storm followed by twenty days stuck in the pack ice. For New Zealand and US visitors, Antarctica is a five-hour flight from Christchurch on a US Air Force C-17 Globemaster, or, when the sea-ice runway isn't operational, an eight-hour flight from Christchurch on a ski-equipped Hercules.

Scott and his men had to pitch tent at the end of each day's work of man-hauling a sledge. Our campsite is at the end of a forty-minute helicopter flight from Scott Base, and all our equipment – fresh food, tents, drilling rig, cameras – was waiting for us when we arrived.

While Scott and his men ate rations of pemmican, biscuits and tea, and finished most meals still hungry, in our field camp we have boxes of "freshies" to select from: we make nachos one night, stir-fried chicken, vegetables and rice the next. In case we get stuck out here, we have also each brought a selection of Back Country Cuisine dehydrated meals, selected from a wide range of options that included nasi goreng, Thai chicken curry and beef stroganoff.

While Scott and his men wore an experimental collection of canvas, wool and reindeer fur, Antarctica New Zealand has supplied each of us with a sophisticated layering system made of lightweight, breathable merino and synthetics. My collection includes four jackets, salopettes, jersey, six pairs of gloves, two pairs of boots, as well as various balaclavas, hats, socks, goggles and thermal underwear. Wearing all of this, including the top layer of ECWs, or Extreme Cold Weather gear, I can expect to keep functioning to something like an insane minus 100°C.

The reindeer-fur sleeping bags of Scott's party became progressively icier and wetter each night from the men's perspiration and condensed breath, so that getting into them was like penetrating a frozen block of ice. In his book *The Worst Journey in the World*, Apsley Cherry-Garrard, who took part in a winter expedition before Scott's attempt on the South Pole, wrote of "the blissful moment of getting out of your bag". Come bedtime I pull on my sleeping mask to shade my eyes from the twenty-four-hour sunlight and snuggle down in my four layers of bedding – a cotton sheet bag, a down sleeping bag, a synthetic sleeping bag and a canvas cover – on my triple-layer sleeping mat: a foam pad, a Thermarest airbed and a sheepskin.

When James Cook completed the first circumnavigation of Antarctica in 1775, it was Terra Australis Incognita, the unknown southern continent. For Scott in the early twentieth century it was The Antarctic. These days, for the scientists and support staff who visit – and some have been going back and forth for decades – it's just The Ice.

After Cook another wave of Antarctic explorers – including James Clark Ross, after whom the Ross Ice Shelf and Ross Island are named – sailed south of the Antarctic Circle without knowing whether or not there was a continent beneath and beyond the icy barrier they encountered. Scott, along with Amundsen, Shackleton, Shirase – the Japanese army officer who led an Antarctic expedition from 1910 to 1912 – and others, came next. Their focus was on "conquering" geographic landmarks such as the south geographic pole, and mysterious convergences such as the south magnetic pole, and claiming them for king and country. Four decades later, during International Geophysical Year, 1957, national bases were established around the fringes of the continent. After an initial focus on geophysical sciences such as meteorology, oceanography and glaciology, there was a proliferation of diverse scientific endeavours, and in 1961, with the signing of the Antarctic Treaty, Antarctica was designated "a continent for science".

I'm not a scientist – I gained a degree in geology but switched to history of science for my PhD topic – but I have spent my entire career involved in science, as a writer, a historian and now a lecturer, teaching science communication, history of science and science writing to mostly science students.

During my 2011 visit to Antarctica I blogged from Scott Base, wrote a series of science articles for *New Zealand Listener*, and collected ideas for this anthology. On the last day I went for a long walk. The last leg of the Castle Rock loop trail took me south-west along the top of Hut Point Peninsula, a thin finger of land that stretches from Mount Erebus to Scott Base and McMurdo Station. From the ridge, visibility was poor. The white curve of the flag-marked path ahead of me quietly merged with the white sky above and the line of the hill slowly disappeared. Looking south over the ice shelf, the white of the ice was mirrored by the white of the cloud layer above, with the only contrast a few black shapes that denoted Willy Airfield. From this distance the loosely related group of shapes – aeroplanes? tractors? buildings? – looked like iron filings scattered on a magnetic table.

Far away in the distance, a narrow arm of pale blue on the horizon marked the thin layer in which we humans, with our buildings, vehicles, aircraft and laboratories, the flotsam and jetsam of science and the military, existed and did our best to survive. The first explorers and scientists mostly explored this layer, gathering rock samples, measuring surface temperatures, and collecting any life forms they came upon. Today scientists are looking beneath the sea ice, often with submersibles, at the algae that live on the undersurface of the ice, at the fish that live beneath the ice, and at the colourful fauna on the sea floor.

Others are probing further still, looking high up into the atmosphere and deep space beyond, bouncing radar into glaciers to see how deep they are, finding and mapping entire mountain ranges buried beneath the ice, uncovering subterranean networks of lakes and rivers. Is there life in these bodies of water beneath the giant ice sheets? The chance is yes – there is life everywhere else people have looked. Extremophiles live in steam vents on Mount Erebus. Other creatures live in small pockets of air in the glaciers of the Dry Valleys. Nematodes lie dormant under rocks, waiting to be revived by drops of water.

Even on my first short visit to Antarctica I got a sense of the breadth of science taking place there. I visited the Adélie penguin colony at Cape Bird, where I was greeted by the unmated male penguins, or "hooligan cocks", that led George Murray Levick to write the privately distributed *The Sexual Habits of the Adélie Penguin*, whose subject matter he deemed unsuitable for non-scientists. I went ice fishing for *Trematomus bernachii* with marine biologist Clive Evans, and met Art DeVries, the American zoologist who identified the protein that lets icefish survive in the minus 2°C waters around the continent.

At a camp on the sea ice I lowered a CTD meter down a hole in the ice and into the ocean to measure how conductivity and temperature changed with depth. I visited the Taylor Valley, and in a Jamesway tent beside Lake Hoare listened to a group of American scientists talk about their research into the Canada Glacier and its meltwaters that form lakes either side of the glacier. And at a "Poking the Pig" scientific lecture at the McMurdo Station Crary Lab I heard twenty-five-year Antarctic veteran Bob Bindschadler talk about his research on the Pine Island Glacier, which he told us – "this sucker really roars" – was one of the fastest moving glaciers on the continent.

But at the end of twelve science-packed days on the ice I felt I'd barely scratched the surface. I was hungry to know more. I read as many scientific accounts of Antarctica as I could find, limited only by the need for them to be written in, or translated into, English. Antarctic scientists recommended topics to explore and books to read. Reading one book led me to another and another. To get an overview of the general shape and scope of Antarctic science I included wide-ranging works such as G.E. Fogg's *A History of Antarctic Science*, Edward Larson's *An Empire of Ice*, and Veronika Meduna's *Science on Ice*.

I made many lists – of scientists whose writings I wanted to include, of key events and discoveries, of iconic species and landscape features – but in the end my selection came down to one thing: Was it a good story?

In the accounts by early explorers and scientists I found pieces about geographical exploration, encounters with new species, and the discoveries of fossil-bearing rocks that showed Antarctic was once a lush green continent. As remote-sensing techniques improved, Antarctic scientists started writing about their explorations of the sub-glacial environment, the atmosphere, and even deep space. As well as accounts of significant science and iconic species, I came upon pieces that documented the everyday business of *doing* Antarctic science. Sometimes the scientific work could be uncomfortable or tedious, as it was for Rhian Salmon one day during her winter-over at Halley Research Station, and sometimes it involved life-threatening challenges, as it did for Richard Byrd when he spent four winter months alone in a hut making meteorological observations. I was also delighted to find some very focused pieces, such as David Campbell on his capture and dissection of a *Euphausia superba*, and Gavin Francis on his investigation of an emperor penguin egg abandoned at a rookery.

The closer I got to contemporary accounts, the more I observed that the science I was finding was about how our planet is changing. Bill Green documented research into the chemistry of streams and lakes in the Dry Valleys and James McClintock revealed how ocean acidification was threatening marine invertebrates such as sea butterflies. I was enthralled with the immediacy of blog posts written by Rob Dunbar, whose sediment drilling projects were revealing what had happened to the Antarctic ice sheets millions of years ago, and by Regina Eisert, whose work on the megafauna of the Ross Sea was providing surprising new evidence on changes occurring to the region's apex predators.

Where I couldn't find a suitable piece on an important topic I commissioned it. I asked Nancy Bertler, whose work I had covered in many magazine articles, to write about her project drilling and examining a 764-metre ice core from Roosevelt Island. And I asked Victoria Metcalf, a marine biologist and science communicator, to write about her decade of research into the bizarre fish that survive and thrive in the sub-zero Antarctic waters.

There are many good books and blogs about Antarctic science written by journalists and other writers, but for this anthology I have chosen only pieces written by the scientists themselves. In some cases these accounts are by people, such as James Cook, James Clark Ross and Robert Falcon Scott, who are commonly thought of as explorers, but from whose voyages came the first reports on many scientific matters, including Antarctic geography and geology and a number of previously unknown species. Most, though, are by people whose primary or sole focus is science: sedimentologists, zoologists, glaciologists, oceanographers, palaeontologists, astrophysicists, and more.

For a different perspective on some of the science, I have included a few poems. Some, such as those by Bill Manhire and Chris Orsman, have been published previously. Others, such as those by Helen Heath and Ashleigh Young, were written in direct response to a piece included in this book. Gregory O'Brien's long poem, which opens the book, reflects on Antarctica as a place that most people will never visit but which looms large in the imagination, particularly for those of us living in the southern hemisphere, and may in time have a profound effect on everyone on the planet.

The pieces in this book are arranged chronologically, under four themes: Unknown Land, The First Antarcticans, Continent for Science and Global Barometer. In the interests of accessibility, silent editorial changes have been made, such as standardising spelling and correcting grammatical errors. Where passages have been deleted, this is clearly shown with either ellipses (...) or a linking editorial passage. Anyone who would like to read the complete version of an excerpted or edited work should refer to the original texts, for which full reference details are given at the end of the book.

After just two days and nights at Friis Hills I start to acclimatise. I remember my Antarctic field training instruction to eat and drink more than usual. I up my fat intake and through Cliff discover the wonders of bags of frozen gummy worms and a thermos filled with hot Raro drink. And I resign myself to the fact that in a blizzard, breathless and on my own in a tent, there is nothing I can do but try to breathe slowly and deeply and marvel at what is likely to be the most remarkable camping experience of my life.

While Adam, the geomorphologist, is convinced the last time the Friis Hills held running water was more than fourteen million years ago, Tim Naish and Richard Levy, the sedimentologists on the team, have discovered another story while working on nearby marine sediments: evidence of a warm period three to five million years ago. Their work on Ross Sea sediment cores during the Andrill geological drilling project revealed that the ocean warmed up, there was no Ross Ice Shelf, the West Antarctic Ice Sheet collapsed, and the ocean was too warm to support sea ice.

What's puzzling is that at the same time the ocean was warming, the mountains seem to have remained cold and frozen. These contradictory pieces of paleoclimate evidence have refuelled a thirty-year debate over what exactly happened to the massive East Antarctic Ice Sheet all those millions of years ago. Did it collapse or remain intact? Sediments that preserve information about land-based climate three to five million years ago are rare, but there are fourteen- to twenty-million-year-old deposits in both marine sediments and Friis Hills.

"That these deposits are so well preserved at Friis Hills implies that much of the high interior East Antarctic Ice Sheet survived the global warmth of the Pliocene, unlike the West Antarctic Ice Sheet, which was attacked by warming oceans," Richard tells me.

While we are at Friis Hills, the team drills into the permafrost to recover a short sediment core to take back to Wellington. We don't need a drill to see the fossils here – the sediments disturbed by the drill crumble easily, revealing the small distinctive leaves of *Nothofagus antarctica*, or southern beech. Using an array of geophones to collect signals from the sound of an enormous hammer banged on to a metal plate, the team also does acoustic surveys across old lake beds to get a more detailed picture of the sub-surface geology. Correlating this land-based evidence with marine drill-core evidence – and determining the likely temperature gradient from mountain to sea – will help reveal what happened to the East Antarctic Ice Sheet during past periods of global warming. The scientists want to know more, and hope to come back in a future field season to drill a deeper hole that may provide information from as far back as twenty million years.

A lot of Antarctic science today is two-pronged: investigating what happened in Antarctica when the world was warmer, and cataloguing the impact that today's climatic and oceanic changes are already having on Antarctic species, seas and ice cover. Determining the conditions under which the East Antarctic Ice Sheet will collapse is important, but even without this we know Antarctica's ice is melting: current estimates are that 120 gigatonnes of ice is being lost each year. At the same time the annual sea ice has reached unprecedented extents, with the 2015 sea ice the biggest on record. What does this mean? There are plenty of theories but only more science can provide the answer.

Not many people get to visit Antarctica: the scientists, base staff, artists, journalists and tourists who visit are a privileged few. But out of sight cannot be out of mind: what happens in this region over the next hundred years will affect us all. At current predictions, average global sea level is due to rise by at least half a metre – from thermal expansion of the ocean as well as melting glaciers and icecaps – by 2100. This is enough to cause serious challenges for much of the world's population, who live in low-lying coastal cities.

The rise will not stop there. Some scientists now believe the entire West Antarctic Ice Sheet may be headed for inevitable collapse, bringing an average sea-level rise across the globe of up to 3.6 metres. If parts of the East Antarctic Ice Sheet were to also collapse – a process likely to take hundreds of years – it could lead to additional sea-level rise of twenty metres.

Those of us fortunate enough to have first-hand experience of Antarctica inevitably come away with profound respect for the place and the people who live and work there. Not only is Earth's last discovered continent a landscape of immense beauty, it is the scene of much vital scientific endeavour. A recent ice sheet model developed by Nick Golledge – a glacial geologist who was part of the Friis Hills team I camped with in 2014 – suggests that if we can keep global warming below 2°C we still have a chance of saving the Antarctic ice sheets from major melting and collapse. That's a goal we can all play a part in achieving.

Rebecca Priestley
January 2016

Unknown Land

Antarctica existed as an idea, an imagined place, long before it was ever sighted. Early mapmakers had suggested there could be a large and populous continent extending from the polar region into the temperate latitudes of the southern hemisphere, to balance the heavily populated European landmass in the northern hemisphere. It was not until the eighteenth century, though, that Europeans ventured far enough south to test the hypothesis that this "terra australis incognita" existed.

These were treacherous journeys in wooden sailing ships, with sailors vulnerable to not just the cold but to malnutrition and diseases such as scurvy on voyages that lasted for years. It was the British Navy captain James Cook who in 1773 led the first expedition to sail inside the Antarctic Circle – the line of latitude, at around 66° South, below which the sun does not set at midsummer.

Once Cook had determined there was indeed an icy expanse below the polar circle, other explorers followed. The focus of these early expeditions was geographical discovery – various parts of Antarctica were "claimed" and named – but their scientific achievements also included documenting and collecting the local flora and fauna, and making observations and measurements of magnetism, astronomy, the aurora australis, wind speed and air temperatures, and variation in the saltiness, currents, temperature and ice conditions of the Southern Ocean. The hunt for the south magnetic pole became a race, with government-sponsored voyages from

Britain, France and the United States setting out to find it. Hunters, too, came south, in search of seals to kill for fur and blubber.

Many Antarctic bases and landforms now bear the names of these early explorers – among them Cook and fellow British Navy captain James Clark Ross, American Navy lieutenant Charles Wilkes and French Navy captain Jules Dumont d'Urville – who skirted around the edge of the continent and on their return published elaborate narratives, which are excerpted in the following pages.

The legacy of the early scientific work also lives on in the species and landforms first named and described – from Mount Erebus, the southernmost volcano in the world, named after one of Ross's ships, to the Adélie penguin named by Dumont d'Urville after his wife.

THE ICE ISLANDS , seen the 9th of Jany. 1773.

Drawn from Nature by W. Hodges.

Engrav'd by B. T. Pouncy.
No XXX.

Drawing by William Hodges, expedition artist on James Cook's second
Pacific voyage, of icebergs in the Southern Ocean, January 9, 1773.

Cook circumnavigates
the continent

"Whether the unexplored part of the southern hemisphere be only an immense mass of water, or contain another continent, as speculative geography seemed to suggest, was a question that had long engaged the attention, not only of learned men, but of most of the maritime powers of Europe," wrote James Cook, a captain in the Royal Navy.

Cook's second Pacific voyage was intended "to put an end to all diversity of opinion about a matter so curious and important". In July 1772, in command of HMS *Resolution* and its 112 men and with Tobias Furneaux in command of HMS *Adventure* and its eighty-one men, Cook sailed from London. The famous sea captain did not find Terra Australis Incognita, a hypothetical continent extending into the southern hemisphere's temperate regions. Nor did he find the Antarctic continent – at his furthest south he was about 100 nautical miles from land. But he did circumnavigate Antarctica, getting as far south as the sea ice, icebergs and ice shelves allowed, and as he left the region he asserted there was a continent beneath and beyond the icy barrier.

As well as looking for land, the men on Cook's voyage measured magnetic variation and dip, observed the stars and the aurora australis, took air and ocean temperatures, and documented variation in the saltiness, currents and ice conditions of the Southern Ocean. The two naturalists on board *Resolution* – father and son Johann and Georg Forster – observed and collected Antarctic and subantarctic wildlife, discovering fourteen seabirds new to science.

In *A Voyage Towards the South Pole and Around the World* Cook documented three legs of the voyage in which his ships skirted the edge of the polar ice. On January 17, 1773, the

Resolution and *Adventure* were the first ships to cross the Antarctic Circle.

In this passage, which starts in November 1773, the *Resolution* leaves from a promontory on the south-east coast of the North Island of New Zealand at an approximate latitude of 41°37' South and longitude 175°17' East. The *Adventure*, after failing to meet the *Resolution* at a pre-arranged place and date, was soon on its way back to Europe.

At eight o'clock in the evening of the 26th we took our departure from Cape Palliser and steered to the south, inclining to the east, having a favourable gale from the NW and SW. We daily saw some rock-weeds, seals, Port Egmont hens, albatrosses, pintadoes, and other petrels, and on the 2nd of December ... we saw a number of red-billed penguins, which remained about us for several days. On the 5th, being in the latitude 50°17' South, longitude 179°40' East, the variation was 18°25' East. At half an hour past eight o'clock the next evening, we reckoned ourselves antipodes to our friends in London, consequently as far removed from them as possible.

On the 8th ... we ceased to see penguins and seals, and concluded that those we had seen retired to the southern parts of New Zealand whenever it was necessary for them to be at land. We had now a strong gale at NW, and a great swell from SW. This swell we got as soon as the south point of New Zealand came in that direction; and as we had had no wind from that quarter the six preceding days, but on the contrary it had been at east, north, and NW, I conclude there can be no land to the southward, under the meridian of New Zealand, but what must lie very far to the south. The two following days we had very stormy weather, sleet and snow, winds between the north and south-west.

The 11th, the storm abated, and the weather clearing up we found the latitude to be 61°15' South, longitude 173°4'W. This fine weather was of short duration: in the evening the wind increased to a strong gale at

SW, blew in squalls attended with thick snow showers, hail, and sleet. The mercury in the thermometer fell to thirty-two; consequently the weather was very cold, and seemed to indicate that ice was not far off.

At four o'clock the next morning ... we saw the first ice island ... At the time we saw this ice we also saw an Antarctic petrel, some grey albatrosses, and our old companions pintadoes and blue petrels. The wind kept veering from SW by the NW to NNE, for the most part a fresh gale, attended with a thick haze and snow; on which account we steered to the SE and E, keeping the wind always on the beam, that it might be in our power to return back nearly on the same track, should our course have been interrupted by any danger whatever. For some days we had a great sea from the NW and SW, so that it is not probable there can be any land near, between these two points.

We fell in with several large islands on the 14th, and, about noon, with a quantity of loose ice, through which we sailed. ... Grey albatrosses, blue petrels, pintadoes, and fulmers were seen. As we advanced to the SE by E with a fresh gale at west, we found the number of ice islands increase fast upon us. Between noon and eight in the evening we saw but two; but before four o'clock in the morning of the 15th, we had passed seventeen, besides a quantity of loose ice which we ran through. At six o'clock, we were obliged to haul to the NE in order to clear an immense field which lay to the south and SE. The ice, in most part of it, lay close packed together; in other places there appeared partitions in the field, and a clear sea beyond it. However, I did not think it safe to venture through, as the wind would not permit us to return the same way that we must go in. Besides, as it blew strong, and the weather at times was exceedingly foggy, it was the more necessary for us to get clear of this loose ice, which is rather more dangerous than the great islands. It was not such ice as is usually found in bays, or rivers, and near shore, but such as breaks off from the islands, and may not improperly be called parings of the large pieces, or the rubbish or fragments which fall off when the great islands break loose from the place where they are formed.

We had not stood long to the NE before we found ourselves embayed by the ice, and were obliged to tack and stretch to the SW having the field, or loose ice, to the south, and many huge islands to the north. After standing two hours on this tack, the wind very luckily veering to the westward, we tacked, stretched to the north, and soon got clear of all the loose ice, but not before we had received several hard knocks from the larger pieces, which, with all our care, we could not avoid. After clearing one danger we still had another to encounter; the weather remained foggy, and many large islands lay in our way, so that we had to luff for one and bear up for another. One we were very near falling aboard of, and if it had happened this circumstance would never have been related. These difficulties, together with the improbability of finding land farther south and the impossibility of exploring it, on account of the ice, if we should find any, determined me to get more to the north. At the time we last tacked, we were in the longitude of 159°20' West and in the latitude of 66°0'S. Several penguins were seen on some of the ice islands, and a few Antarctic petrels on the wing.

We continued to stand to the north, with a fresh gale at west, attended with thick snow showers, till eight o'clock in the evening, when the wind abated, the sky began to clear up, and at six o'clock in the morning of the 16th it fell calm. Four hours after, it was succeeded by a breeze at NE with which we stretched to the SE, having thick hazy weather, with snow showers, and all our rigging coated with ice. In the evening we attempted to take some up out of the sea but were obliged to desist; the sea running too high, and the pieces being so large, that it was dangerous for the boat to come near them.

The next morning, being the 17th, we succeeded better for, falling in with a quantity of loose ice, we hoisted out two boats; and, by noon, got on board as much as we could manage. We then made sail for the east, with a gentle breeze northerly, attended with snow and sleet, which froze to the rigging as it fell. ... The ice we took up proved to be none of the best, being chiefly composed of frozen snow, on which account it was porous and had imbibed a good deal of salt water: but this drained off, after lying a while on deck, and the water then yielded was fresh. We continued to stretch to the east, with a piercing cold northerly wind, attended with a thick fog, snow, and sleet, that decorated all our rigging with icicles. We were hourly meeting with some of the large ice islands, which, in these high latitudes, render navigation so very dangerous: at seven in the evening, falling in with a cluster of them, we narrowly escaped running aboard of one, and with difficulty wore clear of the others. We stood back to the west till ten o'clock, at which time the fog cleared away, and we resumed our course to the east. ...

The clear weather, and the wind, veering to NW, tempted me to steer south; which course we continued till seven in the morning of the 20th when, the wind changing to NE and the sky becoming clouded, we hauled up SE. In the afternoon the wind increased to a strong gale, attended with a thick fog, snow, sleet and rain, which constitutes the very worst of weather. Our rigging, at this time, was so loaded with ice that we had enough to do to get our top-sails down, to double the reef. At seven o'clock in the evening, in the longitude of 147°46', we came, the second time, within the Antarctic or polar circle, continuing our course to the SE till six o'clock the next morning. At that time, being in the latitude of 67°5' South, all at once we got in among a cluster of very large ice islands, and a vast quantity of loose pieces; and as the fog was exceedingly thick, it was with the utmost difficulty we wore clear of them. This done, we stood to the NW till noon when, the fog being somewhat dissipated, we resumed our course again to the SE. The ice islands we met with in the morning were very high and rugged, forming at their tops many peaks, whereas the most of those we had seen before were flat at top, and not so high; though many of them were between two and three hundred feet in height, and between two and three miles in circuit, with perpendicular cliffs or sides, astonishing to behold. Most of our winged companions had now left us; the grey albatrosses only remained, and instead of the other birds we were visited by a few Antarctic petrels.

The 22nd we steered ESE with a fresh gale at north, blowing in squalls, one of which took hold of the mizzen topsail, tore it all to rags, and rendered it, for ever after, useless. At six o'clock in the morning, the wind veering toward the west, our course was east northerly. At this time we were in the latitude of 67°31', the highest we had yet been in, longitude 142°54' West.

We continued our course to the E by N till noon the 23rd, when being in the latitude of 67°12', longitude 138°0', we steered SE; having then twenty-three ice islands in sight from off the deck, and twice that number from the masthead; and yet we could not see above two or three miles round us. At four o'clock in the afternoon ... we fell in with such a quantity of field, or loose ice, as covered the sea in the whole extent from south to east, and was so thick and close as wholly to obstruct our passage. At this time, the wind being pretty moderate, and the sea smooth, we brought to, at the outer edge of the ice, hoisted out two boats and sent them to take some up. In the meantime, we laid hold of several large pieces alongside, and got them on board with our tackle. The taking up ice proved such cold work that it was eight o'clock by the time the boats had made two trips; when we hoisted them in, and made sail to the west, under double-reefed top-sails and courses, with a strong gale at north, attended with snow and sleet, which froze to the rigging as it fell, making the ropes like wires, and the sails like boards or plates of metal. The shivers also were frozen so fast in the blocks that it required our utmost efforts to get a topsail down and up; the cold so intense as hardly to be endured; the whole sea, in a manner, covered with ice; a hard gale, and a thick fog.

Under all these unfavourable circumstances, it was natural for me to think of returning more to the north; seeing no probability of finding any land here nor a possibility of getting farther south. And to have proceeded to the east in this latitude must have been wrong, not only on account of the ice, but because we must have left a vast space of sea to the north unexplored: a space of 24° of latitude, in which a large tract of land might have lain. Whether such a supposition was well-grounded could only be determined by visiting those parts.

While we were taking up ice, we got two of the Antarctic petrels so often mentioned, by which our conjectures were confirmed of their being of the petrel tribe. They are about the size of a large pigeon; the feathers of the head, back, and part of the upper side of the wings are of a light brown; the belly, and underside of the wings, white; the tail feathers are also white, but tipped with brown. At the same time we got another new petrel, smaller than the former and all of a dark grey plumage. We remarked that these birds were fuller of feathers than any we had hitherto seen, such care has nature taken to clothe them suitably to the climate in which they live. At the same time we saw a few chocolate-coloured albatrosses; these, as well as the petrels above mentioned, we nowhere saw but among the ice. Hence one may, with reason, conjecture that there is land to the south. If not, I must ask where these birds breed? A question which perhaps will never be determined, for hitherto we have found these lands, if any, quite inaccessible. ...

On the 24th the wind abated, veering to the NW, and the sky cleared up, in the latitude of 67°0', longitude 138°15'. As we advanced to the NE with a gentle gale at NW, the ice islands increased so fast upon us that this day, at noon, we could see near 100 round us, besides an immense number of small pieces. Perceiving that it was likely to be calm, I got the ship into as clear a berth as I could, where she drifted along with the ice, and by taking the advantage of every light air of wind was kept from falling aboard any of these floating isles. Here it was we spent Christmas day, much in the same manner as we did the preceding one. We were fortunate in having continual daylight and clear weather; for had it been as foggy as on some of the preceding days nothing less than a miracle could have saved us from being dashed to pieces.

> **For most of the next month, the *Resolution* travelled north and then east to escape the ice and explore the seas to the north. In mid January, it started south again. As the narrative resumes it is in calm seas, at latitude 62°34' South, longitude 116°24' West.**

In this situation we had two ice islands in sight, one of which seemed to be as large as any we had seen. It could not be less than 200 feet in height, and terminated in a peak not unlike the cupola of St Paul's church. At this time we had a great westerly swell, which made it improbable that any land should lie between us and the meridian of 133½°, which was our longitude, under the latitude we were now in, when we stood to the north. In all this route we had not seen the least thing that could induce us to think we were ever in the neighbourhood of any land. We had, indeed, frequently seen pieces of seaweed but this, I am well assured, is no sign of the vicinity of land for weed is seen in every part of the ocean. After a few hours calm, we got a wind from SE but it was very unsettled and attended with thick snow showers; at length it fixed at S by E, and we stretched to the east. The wind blew fresh, was piercing cold, and attended with snow and sleet.

On the 22nd, being in the latitude of 62°5' South, longitude 112°24' West, we saw an ice island, an Antarctic petrel, several blue petrels, and some other known birds, but no one thing that gave us the least hopes of finding land.

On the 23rd at noon, we were in the latitude of 62°22'S, longitude 110°24'. In the afternoon we passed an ice island. The wind, which blew fresh, continued to veer to the west, and at eight o'clock the next morning it was to the north of west, when I steered S by W and SSW. … We continued this course till noon the next day, the 25th, when we steered due south. Our latitude, at this time, was 65°24' South, longitude 109°31' West. The wind was at north, the weather mild, and not unpleasant, and not a bit of ice in view. This we thought a little extraordinary as it was but a month before, and not quite 200 leagues to the east, that we were in a manner blocked up with large islands of ice in this very latitude. Saw a single pintadoe petrel, some blue petrels, and a few brown albatrosses. … [The next morning] about noon, seeing the appearance of land to the SE, we immediately trimmed our sails and stood towards it. Soon after it disappeared, but we did not give it up till eight o'clock the next morning, when we were well assured that it was nothing but clouds, or a fog bank; and then we resumed our course to the south, with a gentle breeze at NE, attended with a thick fog, snow and sleet.

We now began to meet with ice islands more frequently than before, and in the latitude of 69°38' South, longitude 108°12' West we fell in with a field of loose ice. As we began to be in want of water, I hoisted out two boats and took up as much as yielded about ten tons. This was cold work but it was now familiar to us. As soon as we had done we hoisted in the boats, and afterwards made short boards over that part of the sea we had, in some measure, made ourselves acquainted with. For we had now so thick a fog that we could not see 200 yards round us; and as we knew not the extent of the loose ice I durst not steer to the south till we had clear weather. Thus we spent the night, or rather that part of the twenty-four hours which answered to night, for we had no darkness but what was occasioned by fogs.

At four o'clock in the morning of the 29th, the fog began to clear away, and the day becoming clear and serene we again steered to the south with a gentle gale at NE and NNE. ... We continued our course to the south and passed a piece of weed covered with barnacles, which a brown albatross was picking off. At ten o'clock we passed a very large ice island; it was not less than three or four miles in circuit. Several more being seen ahead, and the weather becoming foggy, we hauled the wind to the northward, but in less than two hours the weather cleared up and we again stood south.

On the 30th, at four o'clock in the morning, we perceived the clouds over the horizon to the south to be of an unusual snow-white brightness, which we knew denounced our approach to field ice. Soon after it was seen from the top-masthead, and at eight o'clock we were close to its edge. It extended east and west, far beyond the reach of our sight. In the situation we were in just the southern half of our horizon was illuminated, by the rays of light reflected from the ice, to a considerable height. Ninety-seven ice hills were distinctly seen within the field, besides those on the outside; many of them very large, and looking like a ridge of mountains, rising one above another till they were lost in the clouds. The outer or northern edge of this immense field was composed of loose or broken ice close packed together so that it was not possible for anything to enter it. This was about a mile broad, within which was solid ice in one continued compact body. It was rather low and flat (except the hills) but seemed to increase in height as you traced it to the south, in which direction it extended beyond our sight. Such mountains of ice as these were, I believe, never seen in the Greenland Seas, at least not that I ever heard or read of, so that we cannot draw a comparison between the ice here and there. It must be allowed that these prodigious ice mountains must add such additional weight to the ice fields which enclose them as cannot but make a great difference between the navigating this icy sea and that of Greenland.

I will not say it was impossible anywhere to get farther to the south, but the attempting it would have been a dangerous and rash enterprise, and what, I believe, no man in my situation would have thought of. It was, indeed, *my* opinion, as well as the opinion of most on board, that this ice extended quite to the pole, or perhaps joined to some land to which it had been fixed from the earliest time, and that it is here, that is to the south of this parallel, where all the ice we find scattered up and down to the north is first formed, and afterwards broken off by gales of wind, or other causes, and brought to the north by the currents, which we always found to set in that direction in the high latitudes. As we drew near this ice some penguins were heard but none seen, and but few other birds, or any other thing that could induce us to think any land was near. And yet I think there must be some to the south behind this ice. But if there is, it can afford no better retreat for birds, or any other animals, than the ice itself, with which it must be wholly covered. I, who had ambition not only to go farther than anyone had been before but as far as it was possible for man to go, was not sorry at meeting with this interruption as it, in some measure, relieved us, at least shortened the dangers and hardships inseparable from the navigation of the southern polar regions. Since therefore we could not proceed one inch farther to the south, no other reason need be assigned for my tacking and standing back to the north, being at this time in the latitude of 71°10' South, longitude 106°54' West. ...

I was now well satisfied no continent was to be found in this ocean but what must lie so far to the south as to be wholly inaccessible on account of ice, and that, if one should be found in the Southern Atlantic Ocean, it would be necessary to have the whole summer before us to explore it.

From *A Voyage Towards the South Pole and Round the World* by James Cook, volume one: W. Strahan and T. Cadell, London, 1777.

Chinstrap penguins drawn by Pavel Mikhailov, artist on board the
Vostok, on the first Russian Antarctic Expedition, 1819–1821.

Pickled penguins and barrels of ice

Inspired by James Cook's expedition, Thaddeus von Bellingshausen, an officer in the Russian navy, led the ships *Vostok* and *Mirnyi* on a voyage to Antarctica from 1819 to 1821. These would become the second pair of ships – after Cook's – to venture inside the Antarctic Circle.

The main objective – geographical exploration – was achieved. The expedition sighted the ice-covered Antarctic continent and the mountains of the west Antarctic Peninsula, discovered Peter I Island – the first land to be seen within the Antarctic Circle – and sailed through forty-two degrees of longitude while inside the circle, nearly twice as far as Cook.

Bellingshausen's account of the voyage, first published in English translation in 1945, demonstrated just how dependent these early Antarctic explorers were on the scant local resources, such as penguins and icebergs, for food and water.

This narrative starts on January 6, 1820. The ships are at about 60° South and longitude 25° West, approaching Antarctica from the ocean north-east of the Antarctic Peninsula.

At 4 p.m. a fresh wind blew from south by east with snow, so that we made little headway southward. Until 6 p.m. after passing some ice we lay to near a low ice floe on which a great many penguins were sitting. Simanov and Demidov started off in a boat to catch some of them. While they caught some with their hands and stowed them in sacks, the others remained sitting; only a few dived into the water, but without waiting 'til the boat had gone they jumped back on to the floe, helped by the wash of waves.

Our booty consisted of thirty penguins. I ordered a few to be sent to the mess, a few to be prepared for stuffing, and the remainder were kept on board and fed on fresh pork, but this appeared to be injurious to them as they soon sickened, and died after three weeks. The crew skinned them and made caps of the skins, and used the fat for greasing their boots. The penguins were cooked for the officers' mess and we proved that they are good for food, especially if kept for several days in vinegar as is done with certain kinds of game. There was nothing in sight but the unbroken monotony of ice and sea, and the penguin hunt therefore proved a welcome occupation and incidentally provided us with fresh food. We had it stewed together with salt beef and gruel and seasoned with vinegar; the crew liked it, seeing that the officers' mess too pronounced favourably upon it. Fifteen of the penguins were given to the *Mirnyi*.

After the hunt we hoisted the boat and set sail. In the evening when we were in latitude 59°49'50"S, longitude 20°47'W the magnetic variation was found to be 2°34'W. We saw up to twenty-five icebergs and a good deal of broken ice, whilst blue and white petrels and one albatross flew round us continually.

8th. At midnight we had three degrees of frost. We kept to a south-easterly course so as to reach a higher latitude. The wind changed from south by west to south-west and at 3 a.m. the horizon was completely invisible. Between daybreak and 10 a.m. we passed twenty-two icebergs and a large quantity of small broken ice. We approached one of these icebergs on which we could see a great many penguins. We lay to and lowered the boats to cut the ice and to catch as many penguins as we could. Simanov, Lyeskov and Demidov went for the hunt, taking with them part of a fishing net. By midday they had caught thirty-eight. In the meantime the others were cutting ice with which they were able to fill sixteen barrels and all the tubs and cauldrons. We placed the penguins in the chicken runs and in a bath tub placed on the poop for that purpose.

At midday we were in latitude 60°06'08"S, longitude 18°39'51"W. The set of the stream during three days was S.89°E. 39 miles. Owing to heavy snow and a great deal of ice we had to proceed on a different tack until 6 p.m. At this time by observing eight lunar distances we fixed the longitude of our position at 18°12'07"W. The wind had veered to the west. Soon after 6 p.m. we saw on the low flat ice a sea animal; we made for it to see what kind it was and with luck to shoot it. Ignatiev and Demidov, considering themselves fine shots, loaded their guns. The *Mirnyi*, at the same time, went direct towards this ice floe, and as soon as she came within rifle shot an attack was made from both vessels. The animal was wounded in the tail and in two places in the head. The ice was covered with blood. The hunters disputed in a friendly way as to whom the booty should fall, but the dispute remained undecided. Mikhailov drew a sketch of the animal. It was twelve feet long and measured six feet round the body. Its head somewhat resembled a dog's, the tail was short, the upper part of the body was a greenish-grey colour and the underneath was yellow.

Among the crew of the *Vostok* there was a sailor who came from the town of Archangel. He told us that in that district these animals are called "utlyuga". It seemed to be a species of seal. May one conclude, on encountering such animals in the polar seas, that there is land near or not? This question remains unsolved, all the more as these animals may perfectly well breed, change their coats and rest on these ice floes as we saw them. The nearest land known to us was the Sandwich Islands, which were 270 miles away.

> **Over the next two weeks the ships sailed east along the edge of the sea ice, experiencing strong winds, high seas and heavy swells. More dangerous, though, were the frequent snow and thick fog – "the travelling companions of the navigator in the Antarctic" – which made it difficult to see looming icebergs.**
>
> **On January 15 the ships sailed south of latitude 66° South and inside the Antarctic Circle, where they continued to sail**

east, with occasional sightings of petrels, albatrosses, penguins and whales.

26th. The night was clear with stars and there was 3°F of frost. ... In the morning we observed ahead of us, to the east, four large icebergs, and as I had been waiting for some time for fine weather to lay in a supply of ice, I lay to near one of them. The sky was clear and there was a light wind blowing with a slight swell from the south-east, a state of things which rarely occurs in this part of the open ocean. Unless the wind has dropped completely it is impossible to lie alongside an iceberg on account of the surf breaking on the ice. Making use of the favourable weather, we rounded the iceberg to the port side and hove to leeward of the bergs.

Such ice is not always good since, having floated for some time on the surface and being low, it has been continually washed by the waves and becomes friable and contains a certain amount of salt. Nevertheless, in case of necessity, it can be used after it has been treated as follows: Having taken up the ice, leave it for twenty-four hours in the sacks so that all the salt water in it can drain off. In the Antarctic Ocean a great deal of high ice is to be found from which pure ice can be cut which requires no such treatment. The iceberg from which we obtained the ice was 200 feet in height on one side and only 30 feet on the other. In appearance it was smooth with a sloping surface, in length about 125 sazhen [300 yards] and about 60 [140 yards] in width. It was with difficulty that the men could keep a footing on this frozen mass and, as it was both difficult and dangerous to cut the ice at the edges, I decided to give up the attempt and ordered the boats to return and stand by. Meanwhile I had ordered the carronades to be loaded, and

RIGHT: Bellingshausen reported that blue and white petrels flew continually around the Russian ships. These drawings are by the expedition's artist Pavel Mikhailov.

having set sail we turned and stood off the corner of the high ice cliff, reefed the main topsail and then ordered shots to be fired at the very corner. The shock of the shots not only broke off a few fragments of ice, but shook the whole mass of ice to its foundations, so that large pieces fell with a huge crash into the water, raising a vast spray a third of the height of the iceberg itself, and producing for some little while a considerable swell. All this and in addition the sudden appearance of a shoal of whales, which we had not before observed, presented an exceptionally impressive spectacle such as is only possible in the Antarctic. When the iceberg, after some swaying, righted itself, the top of the lower part – that is, the lower side of the iceberg, which had been 30 feet out of the water – was level with the water and the other side was raised correspondingly.

The second lieutenant of the *Vostok*, Lyeskov, celebrated his birthday that day. The captain of the *Mirnyi*, Lazarev, and some of the officers stayed with us until evening. On the suggestion of Lazarev, we fired gun shells from the carronades at the swimming whales. However, partly owing to their very brief stay on the surface of the water and partly to the nature of these guns, which are not well adapted for target practice, our shots missed, falling sometimes wide and sometimes short, and the whales dived, showing their broad horizontal tails. Whalers call this kind of whale "sperm whales" and recognise them solely by their spouting. They reckon that only those which spout twice per minute belong to that species. Whalers were unable to give us any distinguishing features of these whales.

We at once sent the cutter and two boats to cut the ice from the floating blocks. For the quicker despatch of the business, all the crew were set to work. The ice was taken up on both sides and broken into small pieces in a tub (placed for that purpose on the poop), so that it was possible to fill the barrels through the bung holes without having to enlarge the opening. The boats continually brought the ice alongside and, after each trip, the hands at the oars were changed in order to lighten this work, which had to be carried out in the cold and wet. As soon as it was finished all were ordered to change into dry clothing and given a glass of hot punch to refresh them. From 3 p.m. to 10 p.m. forty-nine barrels of medium size were filled with ice, as well as all kettles, cauldrons and a few sacks for immediate use. All the barrels filled up with ice were distributed about on the poop, on the deck and on the forecastle, but none were placed in the hold or on the lower deck, in order to avoid the cold damp air that is given off by the melting of the ice. At last, after hoisting the boats, we turned to the east until three o'clock of the following morning, with a little rain.

From *The Voyage of Captain Bellingshausen to the Antarctic Seas 1819–1821*, by Thaddeus von Bellingshausen, translated from the Russian, edited by Frank Debenham, volume one: Hakluyt Society, London, 1945.

Members of Dumont d'Urville's Antarctic expedition plant the Tricolour and "take possession" of Adélie Land for France, January 21, 1840.

Drinking wine in Adélie Land

France's quest to reach the south magnetic pole was led by Captain Jules Dumont d'Urville. His ship, the *Astrolabe*, was accompanied by the *Zélée* under the command of Charles Jacquinot. If they could not reach the pole, their orders were to get further south than the record of 74°34' South achieved by a British sealer, James Weddell, in 1823.

They did not reach the south magnetic pole, although they made a calculation of its location, but were more successful than a contemporaneous American expedition, led by Charles Wilkes, when it came to finding land. This account is by Joseph Dubouzet, the first lieutenant on the *Zélée*. It is January 21, 1840 and both ships are on or near the polar circle, in the sector of Antarctica directly south of Australia.

Nine days later, the *Astrolabe* would pass very close to and sight an American ship, the *Porpoise*, although neither ship attempted to contact the other.

During the whole day all eyes had been fixed on the coast to try and discover something other than snow and ice. At last, when we were beginning to despair, after having passed a mass of floating islands which quite blocked out the shore, we noticed several little islets the sides of which, destitute of snow, showed us that blackish earth so ardently desired. Several minutes later we saw the large boat of the *Astrolabe* leave the corvette and start towards the shore with an officer and two naturalists. Immediately I requested Commander Jacquinot to embark me in his yawl, which he ordered to be lowered to send ashore.

The *Astrolabe* boat had already got a good start of us; after two and a half hours' hard rowing we reached the nearest islet. Our men were so full of ardour they hardly noticed the effort they had just been making in covering more than seven miles in so short a time. Going along, we passed quite close to immense ice islands. Their perpendicular sides, eaten away by the sea, were crowned on the top by long needles of a greenish ice formed after a thaw. Their appearance was to the last degree imposing. They seemed to form an insurmountable barrier to the east of the islets to which we were bound, and this made me think they were perhaps fixed 80 to 100 fathoms down. Their height indicated about this draught of water. The sea was covered with debris of ice, which obliged us to take a very winding course. On the icebergs (glaçons) we noticed a crowd of penguins, who with a stupid air quietly watched us pass.

It was nearly nine o'clock when, to our great joy, we landed on the western part of the most westerly and the loftiest islet. The *Astrolabe*'s boat had arrived a moment before, and already the men had climbed up the steep sides of this rock. They hurled down the penguins, who were much astonished to find themselves so brutally dispossessed of the island, of which they were the sole inhabitants. We also jumped on shore armed with pickaxes and hammers. The surf rendered this operation very difficult. I was forced to leave several men in the boat to look after her.

I then immediately sent one of our men to unfurl the tricolour flag on this land, which no human creature had either seen or stepped on before. Following the ancient custom, faithfully kept up by the English, we took possession of it in the name of France, as well as of the adjacent coast, which the ice prevented us from approaching. Our enthusiasm and joy were such that it seemed to us we had just added a province to French territory by this wholly pacific conquest. If the abuse that has been born of such acts of possession has caused them to be often regarded as ridiculous and worthless, in this case at any rate we believed ourselves sufficiently in the right to maintain the ancient custom in favour of our country – for we dispossessed none and our titles were incontestable. We regarded ourselves, therefore, at once as being on French soil; and there is at least this advantage that it will never raise up war against our country.

The ceremony ended, as it should, with a libation. To the glory of France, which concerned us deeply just then, we emptied a bottle of the most generous of her wines, which one of our companions had had the presence of mind to bring with him. Never was Bordeaux wine called on to play a more worthy part; never was bottle emptied more fitly. Surrounded on all sides by eternal ice and snow, the cold was extreme. This generous liquor reacted with advantage against the rigours of the temperature. All this happened in less time than it takes to write it. We then all set to work immediately to collect everything of interest in natural history that this barren land could offer. The animal kingdom was only represented by the penguins. Notwithstanding all my search we did not find a single shell. The rock was entirely bare, and did not even offer the least trace of lichens. We found only one single seaweed, and that was dry so it had been brought there by currents or birds.

We were obliged to fall back on the mineral kingdom. Each of us took a hammer and began to hew at the rock. But it was so hard, being of a granite nature, that we could detach only very small pieces. Happily, while wandering on the summit of the island, the sailors discovered large fragments of rock detached by frost, and these they took into our boats. In a short time we had enough to supply specimens to all our museums and to others besides. In examining them closely, I recognised a perfect resemblance between these rocks and some small fragments that we had found in the stomach of a penguin killed the evening before. These fragments could, if necessary, have given an exact idea of the geological formation of this land if it had been impossible to go on shore there. However extraordinary may be this way of doing geology, it proves how much interest the smallest observations may have for the naturalist, often even helping him in his researches by leading him sometimes on to the track of discoveries to which they seem to be the most foreign.

The small islet on which we landed is one of a group of eight or ten small islands, rounded above, and all presenting pretty much the same form. These islands are separated from the nearest coast by a distance of 500 to 600 metres. We noticed along the shore several more tops quite bare, and one cape of which the base was also free from snow, but we noticed also a great quantity of ice, which made the approach to it very difficult. All these islets, very close to each other, seemed to form a continuous chain parallel to the coast from east to west. All the ice islands accumulated in the eastern part, which seemed to me fixed, probably cover other islets similar to those on which we had landed. It is certain that many rocks must be buried every year by enormous masses of ice, of which they form the nucleus. Perhaps even the great land in front of us was cut up by numerous channels. The hydrographical records which were made in these latitudes can have no other object than to determine the form of the glaciers at the moment of our passage without showing the contour of the coast, which must rarely be free from the thick crust covering the soil.

We did not leave these islets till 9.30; we were entranced by the treasures we carried away. Before hoisting sail we saluted our discovery with a general hurrah, to bid it a last goodbye. The echoes of these silent regions, for the first time disturbed by human voices, repeated our cries and then returned to their habitual silence, so gloomy and so imposing. Favoured by a good easterly breeze we took our course to the ships, which were bearing off from land, often disappearing in their tacks behind the great ice islands. We reached them only at 11 p.m. The cold was then extremely sharp. The thermometer registered 5° below zero. The outsides of our boats, as well as the oars, were covered with a coating of ice. We were glad to get back on board the corvettes, happy to have thus completed our discovery without accident, for in this glacial and capricious climate it is not good to leave one's ship for long at a time. The least wind overtaking a ship on such a coast would force it to go out to sea at once and abandon its boats.

This part of the Antarctic continent, visible from the rocky islets where they landed their boats, Dumont d'Urville named Terre Adélie, or Adélie Land, after his wife. The archipelago of rocky islets, which he named Archipel de Pointe-Géologie for the rocks they collected from the site, is now home to the French scientific base Dumont d'Urville Station, which is situated on Île des Pétrels.

Extract by Joseph Dubouzet from "The Journal of M.J. Dumont-D'Urville" in *The Antarctic Manual for the Use of the Expedition of 1901*, edited by George Murray: Royal Geographical Society, London, 1901.

American ship the *Peacock* stuck in Antarctic ice in January 1840, as depicted by Alfred Thomas Agate, the Wilkes' expedition's young artist, alongside an ice shelf described by Wilkes as an "apparently interminable barrier".

Wilkes among the ice islands

While locating the south magnetic pole was a key goal of many scientific expeditions venturing south of the Antarctic Circle, it was not the focus of this 1838–42 American expedition. Instead its leader, Charles Wilkes, was charged with surveying and exploring "the Pacific Ocean and the South Seas" and spent only three months south of latitude 60° South. While they didn't make landfall, the rocks and dirt they found floating on many of the icebergs they encountered was evidence that land was nearby. As Wilkes later wrote, "Much inquiry and a strong desire has been evinced by geologists to ascertain the extent to which these ice islands travel, the boulders and masses of earth they transport, and the direction they take."

Wilkes, a relatively young lieutenant, was in command of what Antarctic historian G.E. Fogg has called an "ill-assorted and shoddily refitted" group of sailing ships, including the three-masted warships *Vincennes* and *Peacock*, two-masted brig *Porpoise*, and smaller pilot boat *Flying Fish*. This excerpt from Wilkes' narrative of the voyage starts in February 14, 1840 when the ships are travelling east along the ice edge in the part of Antarctica that lies south of the west coast of Australia.

14th – At daylight we again made sail for the land, beating in for it until 11 a.m., when we found any further progress quite impossible. I then judged that it was seven or eight miles distant. The day was remarkably clear and the land very distinct. By measurement we made the extent of coast of the Antarctic continent, which was then in sight, seventy-five miles, and by approximate measurement 3,000 feet high. It was entirely covered with snow. Longitude at noon 106°18'42"E, latitude 65°59'40"S, variation 57°05' westerly.

On running in we had passed several icebergs, greatly discoloured with earth, and finding we could not approach the shore any nearer I determined to land on the largest ice island that seemed accessible, to make dip, intensity, and variation observations. On coming up with it, about one-and-a-half miles from where the barrier had stopped us, I hove the ship to, lowered the boats, and fortunately effected a landing. We found embedded in it in places boulders, stones, gravel, sand, and mud or clay. The larger specimens were of red sandstone and basalt. No signs of stratification were to be seen, but it was in places formed of icy conglomerate (if I may use the expression), composed of large pieces of rocks as it were frozen together, and the ice was extremely hard and flint-like. The largest boulder embedded in it was about five or six feet in diameter, but being situated under the shelf of the iceberg we were not able to get at it. Many specimens were obtained, and it was amusing to see the eagerness and desire of all hands to possess themselves of a piece of the Antarctic continent. These pieces were in great demand during the remainder of the cruise.

In the centre of this iceberg was found a pond of most delicious water, over which was a scum of ice about ten inches thick. We obtained from it about 500 gallons. We remained upon this iceberg several hours, and the men amused themselves to their heart's content in sliding. The pond was three feet deep, extending over an area of an acre, and contained sufficient water for half-a-dozen ships. The temperature of the water was 31°. This island had been undoubtedly turned partly over, and had precisely the same appearance the icy barrier would have exhibited if it had been turned bottom up and subsequently much worn by storms. There was no doubt that it had been detached from the land, which was about eight miles distant. ...

Around the iceberg we found many species of zoophytes, viz. salpee, a beautiful specimen of *Clio helicina*, some large pelagie, and many large crustacea. I made several drawings of them. This day, notwithstanding our disappointment in being still repelled from treading on the new continent, was spent with much gratification, and gave us many new specimens from it.

Finding that we had reached the longitude of 105°E before the time anticipated, and being desirous to pursue the discoveries further west, I left a signal flying on this berg, with a bottle containing instructions for the other vessels, directing them to proceed to the westward as far as they could in the time which should remain prior to the 1st of March. At 8 p.m. we joined the ship, and bore away again to the westward, intending to pursue the route pointed out to them.

On the 15th we passed many icebergs, much discoloured with earth, stones, etc., none of which appeared of recent formation. The weather this day became lowering and the breeze fresh; we double-reefed the topsails and made everything snug; the wind was from the southward. At noon we were in longitude 104°E, latitude 64°06'S. The sea had been remarkably smooth the last few days, and no swell, and I began to entertain the idea that we might have a large body of ice to the northward of us, for the position where Cook found the barrier in 1773 was 200 miles further to the north. I determined, however, to pass on in our explorations, hoping they might enable me to join that of Enderby's Land. I deemed it a great object actually to prove the continuity with it if possible. And, if disappointed in this, I should at any rate ascertain whether there had been any change in the ice in this quarter since the time of Cook... .

We had a vast number of whales about us this day, as well as penguins, cape pigeons, white and grey, and small and large petrels. Some seals also were seen. ...

The icebergs were covered with penguins; several officers landed to get a few as specimens. On their return some penguins followed them closely, particularly one, who at last leaped into the boat. It was supposed that its mate had been among those taken and that it had followed on that account. If this were the fact, it would show a remarkable instinctive affection in this bird.

On the 16th the barrier of ice trended to the northward, and we were obliged to haul to the north-east, passing through a large number of ice islands, many of which were stained with earth. In the afternoon a large sea elephant was discovered on the ice; two boats were sent to effect his capture and many balls were fired into him, but he showed the utmost indifference to their effect, doing no more than to raise his head at each shot. He contrived to escape by floundering over the ice until he reached the water, in which he was quite a different being.

At about seven p.m. Dr. Fox was despatched in a boat to visit an ice island that was very much discoloured with clay in patches. He reported that there was upon it a large pond of muddy water, not frozen, although the temperature on board was much below the freezing-point. We observed around the icebergs numerous right whales, puffing in all directions. A large quantity of small crustacea, including shrimps, were here seen around the icebergs. These are believed to be the cause that attracts whales to these parts; they also supply the numerous penguins with their food. For several days I observed a great difference in the wind by day and by night. It had been fresh from the hour of seven in the morning until 8 p.m., when it generally becomes light or dies away altogether. Today we found ourselves in longitude 99°E and latitude 64°21'S. We made observations throughout the twenty-four hours with Leslie's photometer. ...

On the 17th, about 10 a.m., we discovered the barrier extending in a line ahead, and running north and south as far as the eye could reach. Appearances of land were also seen to the south-west, and its trending seemed to be to the northward. We were thus cut off from any further progress to the westward and obliged to retrace our steps. This position of the ice disappointed me, although it concurred with what was reasonably to be expected. We were now in longitude 97°37'E and latitude 64°01'S; our variation was 56°21' westerly, being again on the decrease. Today we had several snow squalls, which, instead of being in flakes, was in small grains, as round as shot, and of various sizes, from that of mustard-seed to buck-shot. It was remarkably dry, pure white, and not at all like hail. We found the bay we had entered was fifty or sixty miles in depth, and having run in on its southern side I determined to return along its northern shore, which we set about with much anxiety, as the weather began to change for the worse. Our situation was by no means such as I should have chosen to encounter bad weather in, the bay being sprinkled with a great many large icebergs. Here we met with a large number of whales, whose curiosity seemed awakened by our presence. Their proximity, however, was anything but pleasant to us, and their blowings resembled that of a number of locomotives. Their close approach was a convincing proof that they had never been exposed to the pursuit of their skilful hunters. They were of the fin-back species and of extraordinary size.

Between ten and eleven o'clock at night it was entirely clear overhead, and we were gratified with a splendid exhibition of the aurora australis. It exceeded anything of the kind I had heretofore witnessed. Its activity was inconceivable, darting from the zenith to the horizon in all directions in the most brilliant coruscations: rays proceeding as if from a point in the zenith, flashed in brilliant pencillings of light, like sparks of electric fluid *in vacuo*, and reappear again to vanish; forming themselves into one body, like an umbrella, or fan, shut up; again emerging to flit across the sky with the rapidity of light, they showed all the prismatic colours at once or in quick succession. So remarkable were the phenomena that even our sailors were constantly exclaiming in admiration of its brilliancy. The best position in which to view it was by lying flat upon the deck and looking up. The electrometer was tried but no effect perceived. The star Canopus was in the zenith at the time, and though visible through the aurora was much diminished in brightness. On this night also the moon was partially eclipsed.

Large icebergs had now become very numerous, and strengthened the belief that the land existing in this vicinity had taken a very decided trend to the northward. I accordingly followed up the northern barrier closely and passed through the thickest of these bergs, well knowing from our experience that we should have little or no opportunity of seeing the land unless on the inner side of them. It appeared as though they had collected here from other places, and it is impossible to form an idea of the small space to which we were at times confined. Upwards of 100 ice islands, some of which were several miles long, could be counted at a time without the aid of a glass. We enjoyed this beautiful sight with the more pleasure for we had become used to them and knew from experience it was possible to navigate through them without accident.

On the 18th we continued beating to the eastward, and found no end to the apparently interminable barrier. We had a smooth sea and better weather than I anticipated. At noon we had retraced our way about forty miles. Today we again had snow, which fell in the form of regular six-pointed stars. The needles of which these stars were formed were quite distinct and of regular crystals. The temperature at the time was 28°. The barometer stood at 28.76 inches, about three-tenths lower than we had had it for the last twelve days. The wind was easterly.

19th – During this day the barrier trended more to the north-east, and we not unfrequently entered bays so deep as to find ourselves, on reaching the extremity, cut off by the barrier and compelled to return to within a few miles of the place where we had entered. I thought at first that this might have been caused by the tide or current, but repeated trials showed none. Neither did I detect any motion in the floating ice except what was caused by the wind. Our longitude today was 101°E, latitude 63°02'S. Some anxiety seemed to exist among the officers and crew lest we should find ourselves embayed or cut off from the clear sea by a line of barrier. There appeared strong reason for this apprehension, as the smooth sea we had had for several days still continued, we had been sailing as if upon a river, and the water had not assumed its blue colour.

It was, therefore, with great pleasure that on the 20th a slight swell was perceived, and the barrier began to trend more to the northward, and afterwards again to the westward. In the morning we found ourselves still surrounded by great numbers of ice islands. After obtaining a tolerably clear space, the day being rather favourable, we sounded with the deep-sea line 850 fathoms; Six's thermometer gave at the surface 31°, and at the depth of 850 fathoms 35°, an increase of four degrees. The current was again tried but none was found. A white object was visible at eleven fathoms. The water had now assumed a bluish cast.

We endeavoured today to land on an iceberg, but there was too much sea. Shrimps were in great quantities about it but swam too deep to be taken. The wind again hauled to the westward, which disappointed me as I was in hopes of getting to the position where Cook saw the ice in 1773, being now nearly in the same latitude. It was less than 100 miles to the westward of us, and little doubt can exist that its situation has not changed materially in sixty-seven years. ...

In some places we sailed for more than 50 miles together, along a straight and perpendicular wall, from 150 to 200 feet in height, with the land behind it. The icebergs found along the coast afloat were from a quarter of a mile to five miles in length; their separation from the land may be effected by severe frost rending them asunder, after which the violent and frequent storms may be considered a sufficient cause to overcome the attraction which holds them to the parent mass. In their next stage they exhibit the process of decay, being found 50 or 60 miles from the land, and for the most part with their surfaces inclined at a considerable angle to the horizon. This is caused by a change in the position of the centre of gravity, arising from the abrading action of the waves.

By our observations on the temperature of the sea, it is evident that these ice islands can be little changed by the melting process before they reach the latitude of 60°. The temperature of the sea (as observed by the vessels going to and returning from the south) showed but little change above this latitude, and no doubt it was at its maximum as it was then the height of the summer season.

During their drift to the northward, reaching lower latitudes and as their distance from the land increases, they are found in all stages of decay: some forming obelisks; others towers and Gothic arches; and all more or less perforated. Some exhibit lofty columns, with a natural bridge resting on them, of a lightness and beauty inconceivable in any other material. ...

While in this state they rarely exhibit any signs of stratification and some appear to be formed of a soft and porous ice; others are quite blue; others again show a green tint, and are of hard flinty ice. Large ice islands are seen that retain their tabular tops nearly entire until they reach a low latitude, when their dissolution rapidly ensues; while some have lost all resemblance to their original formation and had evidently been overturned.

The process of actually rending asunder was not witnessed by any of the vessels, although in the *Flying Fish*, when during fogs they were in close proximity to large ice islands, they inferred from the loud crashing and the sudden splashing of the sea on her that such occurrences had taken place. As the bergs gradually become worn by the abrasion of the sea they in many cases form large overhanging shelves, about two or three feet above the water, extending out ten or twelve feet. The under part of this projecting mass exhibits the appearance of a collection of icicles hanging from it. The temperature of the water, when among the icebergs, was found below or about the freezing point.

I have before spoken of the boulders embedded in the icebergs. All those that I had an opportunity of observing, apparently formed a part of the nucleus, and were surrounded by extremely compact ice, so that they appear to be connected with that portion of the ice that would be the last to dissolve, and these boulders would therefore, in all probability be carried to the farthest extent of their range before they were let loose or deposited.

On February 21, after nearly six weeks navigating their way through fields of icebergs, Wilkes made the decision to leave the Antarctic and head north.

I have seldom seen so many happy faces, or such rejoicings, as the announcement of my intention to return produced. But although the crew were delighted at the termination of this dangerous cruise, not a word of impatience or discontent had been heard during its continuance. Neither had there been occasion for punishment; and I could not but be thankful to have been enabled to conduct the ship through so difficult and dangerous a navigation without a single accident, with a crew in as good if not in better condition than when we first reached the icy barrier. For myself, I indeed felt worse for the fatigues and anxieties I had undergone but I was able to attend to all my duties, and considered myself amply repaid for my impaired health by the important discoveries we had made, and the success that had attended our exertions.

From *Narrative of the United States Exploring Expedition, During the years 1838, 1839, 1840, 1841, 1842* by Charles Wilkes, volume one: Ingram, Cooke, and Co, London, 1852.

British ships *Erebus* and *Terror* lying off Ross Island, with the active volcano now known as Mount Erebus in the distance.

Fire and ice

James Clark Ross led a British voyage of scientific discovery
to Antarctica and the Southern Ocean from 1839 to 1843. A
Royal Navy captain and veteran of fifteen summers and eight
winters in the Arctic, Ross had been part of a sledging team
that in 1831 located the north magnetic pole in the Canadian
Arctic. His voyage south was tasked with collecting botanical,
zoological and geological samples, and making meteorological
and oceanographic observations, but its prime purpose was
magnetic research.

The expedition's officers took magnetic observations –
declination, dip, and intensity – along their journey, and set
up three southern hemisphere fixed magnetic observatories.
Their orders were to then head south "to determine the position
of the magnetic pole, and even to attain to it if possible". In
the event of their finding any significant extent of land, they
were to attempt to map the coastline and sail as far south as
was possible.

The story opens in January 1841, on Ross's first venture
south of the polar circle. The voyage's two reinforced sailing
ships – HMS *Erebus* under Ross's command and HMS *Terror*
with Commander Francis Crozier at the helm – are just north
of what is now called Ross Island, in McMurdo Sound.

At between two and three miles distance from the land the soundings
were regular, in thirty-eight to forty-one fathoms, on a bed of fine
sand and black stones, and probably good anchorage might be found
near the shore with southerly winds. A high cliff of ice projects into
the sea from the south and south-west sides, rendering it there quite
inaccessible, and a dangerous reef of rocks extends from its southern
cape at least four or five miles, with apparently a deep water passage

between them and the cape; several icebergs of moderate size were aground on the banks to the northward and westward of the island. At midnight the bearings of eight separate islands are given in the log of the *Erebus*, but as these afterwards proved to be the summits of mountains, at a great distance, belonging to the mainland, they do not appear upon the chart as islands. With a favourable breeze and very clear weather, we stood to the southward, close to some land which had been in sight since the preceding noon, and which we then called the "High Island". It proved to be a mountain 12,400 feet of elevation above the level of the sea, emitting flame and smoke in great profusion. At first the smoke appeared like snow drift, but as we drew nearer its true character became manifest.

The discovery of an active volcano in so high a southern latitude cannot but be esteemed a circumstance of high geological importance and interest, and contribute to throw some further light on the physical construction of our globe. I named it Mount Erebus, and an extinct volcano to the eastward, little inferior in height, being by measurement 10,900 feet high, was called Mount Terror.

A small high round island, which had been in sight all the morning, was named Beaufort Island, in compliment to Captain Francis Beaufort of the Royal Navy, Hydrographer to the Admiralty, who was not only mainly instrumental in promoting the sending forth of our expedition, but afforded me much assistance during its equipment by his opinion and advice, and it is very gratifying to me to pay this tribute of respect and gratitude to him for the many acts of kindness and personal friendship I have received at his hands. At 4 p.m. we were in latitude 76°6'S, longitude 168°11'E. The magnetic dip was 88°27'S and the variation 95°31': we were therefore considerably to the southward of the magnetic pole, without any appearance of being able to approach it on account of the land-ice, at a short distance to the westward, uniting with the western point of the "High Island", which, however, afterwards proved to be part of the main land and of which Mount Erebus forms the most conspicuous object.

As we approached the land under all studding-sails, we perceived a low white line extending from its eastern extreme point as far as the eye could discern to the eastward. It presented an extraordinary appearance, gradually increasing in height as we got nearer to it, and proving at length to be a perpendicular cliff of ice, between 150 and 200 feet above the level of the sea, perfectly flat and level at the top, and without any fissures or promontories on its even seaward face. What was beyond it we could not imagine, for being much higher than our masthead, we could not see anything except the summit of a lofty range of mountains extending to the southward as far as the seventy-ninth degree of latitude. These mountains, being the southernmost land hitherto discovered, I felt great satisfaction in naming after Captain Sir William Edward Parry, R.N., in grateful remembrance of the honour he conferred on me, by calling the northernmost known land on the globe by my name; and more especially for the encouragement, assistance and friendship which he bestowed on me during the many years I had the honour and happiness to serve under his distinguished command on four successive voyages to the Arctic seas; and to which I mainly attribute the opportunity now afforded me of thus expressing how deeply I feel myself indebted to his assistance and example.

Whether Parry Mountains again take an easterly trending, and form the base to which this extraordinary mass of ice is attached, must be left for future navigators to determine. If there be land to the southward it must be very remote, or of much less elevation than any other part of the coast we have seen, or it would have appeared above the barrier. Meeting with such an obstruction was a great disappointment to us all, for we had already, in expectation, passed far beyond the eightieth degree, and had even appointed a rendezvous there, in case of the ships accidentally separating. It was, however, an obstruction of

Erebus and *Terror* alongside what is now known as the Ross Ice Shelf. Ross wrote: "Meeting with such an obstruction was a great disappointment ... we might with equal chance of success try to sail through the cliffs of Dover as penetrate such a mass."

such a character as to leave no doubt upon my mind as to our future proceedings, for we might with equal chance of success try to sail through the Cliffs of Dover as penetrate such a mass.

When within three or four miles of this most remarkable object, we altered our course to the eastward for the purpose of determining its extent, and not without the hope that it might still lead us much further to the southward. The whole coast here from the western extreme point now presented a similar vertical cliff of ice, about two or three hundred feet high. The eastern cape at the foot of Mount Terror was named after my friend and colleague Commander Francis Rawdon Moira Crozier of the *Terror*, to whose zeal and cordial cooperation is mainly to be ascribed, under God's blessing, the happiness as well as success of the expedition. Under the circumstances we were placed in, it is impossible for others fully to understand the value of having so tried a friend, of now more than twenty years' standing, as commander of the second ship, upon whom the harmony and right feeling between the two vessels so greatly depends. I considered myself equally fortunate in having for the senior lieutenant of the *Erebus* one whose worth was so well known to me, and who, as well as Commander Crozier, had ever shown so much firmness and prudence during the arduous voyages to the Arctic regions, in which we sailed together as messmates, under the most successful Arctic navigator. In compliment to him I named the western promontory at the foot of Mount Erebus Cape Bird. These two points form the only conspicuous headlands of the coast, the bay between them being of inconsiderable depth.

At 4 p.m. Mount Erebus was observed to emit smoke and flame in unusual quantities, producing a most grand spectacle. A volume of dense smoke was projected at each successive jet with great force, in a vertical column to the height of between 1,500 and 2,000 feet above the mouth of the crater, when, condensing first at its upper part, it descended in mist or snow and gradually dispersed, to be succeeded by another splendid exhibition of the same kind about half an hour afterwards, although the intervals between the eruptions were by no means regular. The diameter of the columns of smoke was between two and three hundred feet as near as we could measure it; whenever the smoke cleared away, the bright red flame that filled the mouth of the crater was clearly perceptible, and some of the officers believed they could see streams of lava pouring down its sides until lost beneath the snow which descended from a few hundred feet below the crater, and projected its perpendicular icy cliff several miles into the ocean. Mount Terror was much more free from snow, especially on its eastern side, where were numerous little conical crater-like hillocks, each of which had probably been, at some period, an active volcano; two very conspicuous hills of this kind were observed close to Cape Crozier. The land upon which Mounts Erebus and Terror stand, comprised between Cape Crozier and Cape Bird, had the appearance of an island from our present position, but the fixed ice, not admitting of our getting to the westward of Cape Bird, prevented our ascertaining whether it was so or not at this time.

> **Erebus is the southernmost active volcano on Earth, and in a near constant state of low-level eruption. The New Zealand and United States scientific bases, Scott Base and McMurdo Station, are now located some 35 kilometres south-west of Mount Erebus on Ross Island.**

From *A Voyage of Discovery and Research in the Southern and Antarctic Regions during the Years 1839–43* by James Clark Ross, volume one: John Murray, London, 1847.

The polar captain's wife

I imagine you are much
preoccupied with the cold,

inching your way between
icebergs of a menacing blue.

This picture satisfies us both;
it is your duty to sound out

the buoyant and laden,
and think betimes of your wife

stranded amidst the furniture.
Being a little drunk I wander

from room to room, touching
temperate surfaces –

a book, a clock, a chair –
missing the body heat

we stowed in that chamber
our last night together.

Chris Orsman

The First Antarcticans

The next time Antarctica became a focus of international attention was in the early years of the twentieth century. One newly interested group were the whalers, who – now that northern hemisphere whales were becoming scarce – headed to Antarctica with their fast boats and explosive harpoons. They would hunt the southern hemisphere whales almost to extinction.

The other group, hardy explorers, went with the aim of reaching the south geographic pole. Antarctica is covered in ice, and at the beginning of what is now called the heroic age no one even knew if there was a continent there, or just a series of islands emerging from the ice. Expeditions were launched from Belgium, France, Norway, Germany, Russia, Sweden, Britain, Japan and Australasia. Intent on reaching land to set up bases, many of the ships became stuck in sea ice, either fated to spend a year or more in Antarctica before being freed – in 1898, the *Belgica* was the first ship to spend the winter frozen into the ice – or doomed, like Shackleton's *Endurance*, which was crushed by sea ice and sank in 1915.

The era is known for such heroic pursuits as Roald Amundsen's conquest of the South Pole, Robert Falcon Scott's death, with his men, on his return from the pole, and Ernest Shackleton's epic 1915 journey from Antarctica to South Georgia. But this was also a period of significant science. After the South Pole was reached in 1911 exploration continued, with inland mountains, glaciers, and the remarkable McMurdo

Dry Valleys being explored and named. Explorers and scientists experimented with different ways of travelling in this icy continent – first with dogs and ponies, later with tractors and planes – but many of the early journeys involved man-hauling. In this mode of transport, men travelled on skis, dragging behind them heavy sledges loaded with not just what they needed to survive but the tools they needed to do their science. Adding to their loads were samples to take back to their (mostly) European homes, from fossil-bearing rocks, which revealed that Antarctica had once had a warmer climate, to emperor penguin eggs, acquired after a deathly cold expedition in the dark of the Antarctic winter.

These expeditions published multi-volume illustrated reports, detailing Antarctic biology, zoology, botany, glaciology, geography, meteorology, hydrology and more. The discoveries of later scientists might contradict some of the findings – Scott, for example, falsely declared the Dry Valleys devoid of life – but the era laid the groundwork for much of the twentieth-century science that followed.

The aurora australis over Antarctica, described by Frederick Cook, a member of the Belgian Antarctic Expedition of 1897–99, as "a trembling lace-work, draped like a curtain, on the southern sky".

In a sleeping bag beneath the aurora australis

The first ship to winter over in Antarctica was RV *Belgica* in 1898. This Belgian Antarctic Expedition was led by Adrien de Gerlache, an officer in the Belgian Royal Navy, and had a multinational crew, including Norwegian Roald Amundsen as first mate and American Frederick Cook as expedition surgeon, anthropologist and photographer.

The *Belgica* crossed the Antarctic Circle on February 15 and later that month became trapped in pack ice in the Bellingshausen Sea, west of the Antarctic Peninsula. During the twelve months the ship was stuck the eighteen men on board made pioneering observations of Antarctic meteorology, geology and zoology. Fresh seal and penguin meat – which Frederick Cook hunted and insisted they eat – helped protect them from scurvy and even starvation.

In this account written in March, Cook describes his first sighting of the aurora australis. The crews of earlier expeditions had seen the aurora, which was believed to be associated with Earth's magnetic field, but the long dark winter provided many more viewing opportunities.

Last night was clear and blue. We knew from the stillness of the air and crackle of the ice that it would be very cold and so it proved. At six o'clock it was −14.6°C, at midnight −20°C. A number of royal and small penguins and some seals were led by curiosity to visit us. They called, and cried, and talked, and grunted, as they walked over the ice about the ship, and were finally captured by the naturalist and the cook, who had an equal interest in the entertainment of our animal friends and in their future destiny.

A few nights past a sea leopard interviewed the meteorologist, Arctowski. The animal sprang suddenly from a new break in the ice on to the floe, upon which Arctowski had a number of delicate meteorological instruments, and without an introduction or any signs of friendship the animal crept rapidly over the snow and examined Arctowski and his paraphernalia with characteristic seal inquisitiveness. The meteorologist had nothing with which to defend himself, and he didn't appear to relish the teeth of the leopard as it advanced and separated its massive jaws with a bear-like snort. He walked around the floe, the leopard after him. The seal examined the instruments, but they were not to its liking, and as to Arctowski, it evidently did not regard him of sufficient interest to follow long, for after it had made two rounds the seal plunged into the waters, swam under the ice and around the floe, and then raised its head far out to get another glimpse of the meteorologist. Thinking that the creature contemplated another attack, Arctowski made warlike gestures and uttered a volley of sulphurous Polish words, but the seal didn't mind that. It raised its head higher and higher out of the water, and displayed its teeth in the best possible manner. Now and then its lips moved and there was audible a weird noise, with signs which we took to be the animal's manner of inviting its new acquaintance to a journey under the icy surface, where they might talk over the matter out of the cold blast of the wind, in the blue depths below.

March 15 – The weather is remarkably clear. There is no wind, no noise, and no motion in the ice. During the night we saw the first aurora australis. I saw it first at eight o'clock, but it was so faint then that I could not be positively certain whether it was a cloud with an unusual ice-blink upon it or an aurora, but at ten o'clock we all saw it in a manner which was unmistakable. The first phenomenon was like a series of wavy fragments of cirrus clouds, blown by strong high winds across the zenith. This entirely disappeared a few minutes after eight o'clock. What we saw later was a trembling lace-work, draped like a curtain, on the southern sky. Various parts were now dark, and now light, as if a stream of electric sparks illuminated the fabric. The curtain seemed to move in response to these waves of light, as if driven by the wind which shook out old folds and created new ones, all of which made the scene one of new interest and rare glory.

That I might better see the new attraction and also experiment with my sleeping bag, I resolved to try a sleep outside upon one of the floes. For several days I had promised myself the pleasure of this experience, but for one reason or another I had deferred it. At midnight I took my bag and, leaving the warmth and comfort of the cabin, I struggled out over the icy walls of the bark's embankment, and upon a floe 300 yards east I spread out the bag. The temperature of the cabin was the ordinary temperature of a comfortable room; the temperature of the outside air was –20°C. After undressing quickly, as one is apt to do in such temperatures, I slid into the fur bag and rolled over the ice until I found a depression suitable to my ideas of comfort. At first my teeth chattered and every muscle of my body quivered, but in a few minutes this passed off and there came a reaction similar to that after a cold bath. With this warm glow I turned from side to side and peeped past the fringe of accumulating frost, around my blowhole through the bag, at the cold glitter of the stars. As I lay there alone, away from the noise of the ship, the silence and the solitude were curiously oppressive. There was not a breath of air stirring the glassy atmosphere, and not a sound from the ice-decked sea or its life to indicate movement or commotion. Only a day ago this same ice was a mass of small detached floes, moving and grinding off edges with a complaining squeak. How different it was now! Every fragment was cemented together into one heterogeneous mass and carpeted by a hard ivory-like sheet of snow. Every move which I made in my bag was followed by a crackling complaint from the snow crust.

At about three o'clock in the morning a little wind came from the east. My blowhole was turned in this direction, but the slow blast of air which struck my face kept my moustache and my whiskers, and every bit of fur near the opening, covered with ice. As I rolled over to face the leeward there seemed to be a misfit somewhere. The hood portion of the bag was as hard as if coated with sheet iron, and my head was firmly encased. My hair, my face, and the undergarments about my neck were frozen to the hood. With every turn I endured an agony of hair pulling. If I remained still my head became more and more fixed by the increasing condensation. In the morning my head was boxed like that of a deep-sea diver. But aside from this little discomfort I was perfectly at ease, and might have slept if the glory of the heavens and the charm of the scene about had not been too fascinating to permit restful repose.

The aurora, as the blue twilight announced the dawn, had settled into an arc of steady brilliancy which hung low on the southern sky, while directly under the zenith there quivered a few streamers; overhead was the southern cross, and all around the blue dome there were sparkling spots which stood out like huge gems. Along the horizon from south to east there was the glow of the sun, probably reflected from the unknown southern lands. This was a band of ochre tapering to gold and ending in orange red.

At four o'clock the aurora was still visible but faint. The heavens were violet and the stars were now fading behind the increasing twilight. A zone of yellow extended from west around south to east, while the other half of the circle was a vivid purple. The ice was a dark blue. An hour later the highest icebergs began to glitter as if tipped with gold, and then the hummocks brightened. Finally, as the sun rose from her snowy bed, the whole frigid sea was coloured as if flooded with liquid gold. I turned over and had dropped into another slumber when I felt a peculiar tapping on the encasement of my face. I remained quiet, and presently I heard a loud chatter. It was uttered by a group of penguins who had come to interview their new companion. I hastened to respond to the call, and after pounding my head and pulling out some bunches of hair I jumped into my furs, bid the surprised penguins good morning, and went aboard. Here I learned that Lecointe, not knowing of my presence on the ice, had taken me for a seal, and was only waiting for better light to try his luck with the rifle.

That winter de Gerlache and Lecointe, the ship's captain, became incapacitated with scurvy and Cook and Amundsen took command of the *Belgica*. When summer arrived the ship was able to escape from the ice and return to Belgium.

Some ten years later, Frederick Cook set off with two Inuit men on an expedition to the North Pole. His claim to have reached the Pole in April 1908 was widely disputed, as was that of fellow American Robert Peary a year later. Roald Amundsen had been planning a trip to the North Pole but after hearing Cook's and Peary's claims he decided on a more ambitious and ultimately successful mission to be the first to reach the South Pole.

From *Through the First Antarctic Night, 1898–1899* by Frederick A. Cook: DoubleDay, Page & Company, New York, 1909.

Members of Robert Falcon Scott's British Antarctic Expedition of 1910–13 man-hauling a sledge near the Beardmore Glacier.

The art and science of sledge travel

"No polar vessel ever left these shores so well adapted and prepared as the *Discovery*," Clements Markham wrote in the preface to *The Antarctic Manual*, a collection of articles and book excerpts on polar science and discovery produced to assist the British Antarctic Expedition of 1901–04. Markham was president of the Royal Geographical Society and a driving force behind the expedition.

Most pieces in the manual were specifically about Antarctica, but this one about sledging was by veteran Arctic explorer Leopold McClintock. The *Discovery* team was planning journeys by sledge to explore the Ross Ice Shelf and Polar Plateau, and the advice from McClintock, who had pioneered long-distance Arctic sledge travelling and devised a system of man-hauling sledges that differed from the traditional dog-pulled sledge, was welcomed by the *Discovery* team. His paper had first appeared in the Royal Geographical Society's *Proceedings* of 1875.

The tent requires but little description. It is merely a pent roof, about seven feet high along the ridge, supported on boarding pikes or poles, crossed at each end, and covering an oblong space sufficient to enclose the party when lying down, and closely packed together. Its duty is merely to afford shelter from the wind and snowdrift, and its weight, when completely fitted, is, for a party of eight men, only about forty pounds. It is made of light closely woven duck.

The sledge is a much more important article of equipment. That which our experience has proved to be the most suitable is a large runner sledge.

It must be borne in mind that I am speaking of latitudes beyond the

70th parallel, where, unlike regions which lie somewhat less remote, the fall of snow is less considerable and never deep; and, moreover, that our sledges often have to be drawn over the sea ice when flooded with water a foot in depth.

The runners are rather broad – that is, three inches – and they stand high, carrying the lading about a foot above the ice. An average-sized sledge is three feet wide and ten feet long, and is drawn by seven men. It is constructed with only just so much strength as is absolutely necessary, since every pound of weight saved in wood and iron enables so much more provisions to be carried. All our sledges have been drawn by the seamen, and the labour of doing so is most excessive. ...

It is necessary to apprehend clearly the nature of the surface over which our sledges have to travel. People unacquainted with the subject commonly fall into one or other extreme, and suppose that we either skate over glassy ice or walk on snowshoes over snow of any conceivable depth.

Salt-water ice is not so smooth as to be slippery; to skate upon it is very possible, though very fatiguing. But hardly is the sea frozen over when the snow falls and remains upon it all the winter. When it first falls the snow is soft and perhaps a foot or fifteen inches deep, but it is blown about by every wind until having become like the finest sand, and hardened under a severe temperature, it consolidates into a covering of a few inches in depth, and becomes so compact that the sledge runner does not sink more than an inch or so: its specific gravity is then about half that of water.

This expanse of snow is rarely smooth: its surface is broken into ridges or furrows by every strong wind. These ridges are the "sastrugi" of Admiral Wrangell, and although the inequalities are seldom more than a foot high they add greatly to the labour of travelling, especially when obliged to cross them at right angles.

As the spring season advances the old winter snow becomes softened, fresh snow falls, and sledging is made more laborious still. At length the thaw arrives; the snow becomes a sludgy mixture, with wet snow on top and water beneath, through which men and sledges sink down to the ice below. It is now almost impossible to get along at all, but in a few days the snow dissolves and we make fair progress again over the now flooded ice.

Our dry provisions and clothing are so packed upon the sledges as to be protected from the wet, but everyone is of course drenched, and remains so during the march through this ice-cold water. This is cold-water cure in real earnest, but I would not recommend any one with the slightest suspicion of a rheumatic tendency to try it!

Later still the water drains off the sea ice through cracks or holes decayed in it, and only tortuous pools of water remain upon it. Later than this, sledge travelling, without the accompaniment of a boat, becomes unsafe.

Such is the nature of the travelling when the sea ice has not been crushed up into hummocks, or masses of various sizes and shapes.

We seldom find either unbroken ice, or ice so crushed up into ridges that we cannot get over it at all, but as a rule crushed up or hummocky ice, three or four feet in height, is of very frequent occurrence, and of course adds much to the labour of sledging. ...

The provisions and the clothing found to be most suitable may now be briefly described. Tea, chocolate, biscuit, preserved meat and pemmican are commonly used. Pemmican is a description of preserved meat used by the Indians of North America, from whom it has been copied. It is a preparation of beef, whereby all that is fluid is evaporated over a wood fire; the fibre is then pounded up and mixed with an equal weight of melted beef fat; no salt or preservative of any kind is used. No more concentrated food for working men in a cold climate is known.

With chocolate, biscuit and a little warmed-up pemmican, the traveller makes a good breakfast. A few ounces of specially prepared bacon, almost free from salt, some biscuit and a mouthful of grog forms his hasty luncheon on the march. And, on encamping, he has his supper of warmed pemmican, or other preserved meat, and tea.

Rum is the spirit used in the navy, and therefore in our Arctic ships. If the men were not accustomed to the use of spirits I think that, except on special occasions when a stimulant is desirable, they would be even better without it as an equal weight of some nutritious food might then be carried instead. However, the ration of rum is very small.

This simple diet is invariable, except when the party is so fortunate as to procure game, and then the awkward question crops up of fuel wherewith to cook it. We are at a disadvantage with those hardy men who are content to cook their meat with frost, although a sandwich of frozen bear's blubber and biscuit is palatable enough, and I think most of the gentlemen in this room would agree with me if they were fairly educated up to it by a few days' sledging in the month of March.

All our cooking is done with lamps, the fuel being either spirits of wine, or some fatty substance such as stearine of cocoa-nut oil, tallow or blubber. The latter alone is used by the Esquimaux. We prefer the stearine as it cooks more rapidly and makes less smoke, and the stearine lamp suits equally well for blubber, or any animal fat procurable on the march.

The clothing of the men is a subject of equal difficulty and importance. It must be suited to the temperature under which they travel, and this often ranges over 100 degrees – that is, from +50° to −50°; it must not suffer by frequent wettings and should dry quickly; and, as only the outer wrappings of the feet are ever taken off while the frost lasts, it should also be suitable for sleeping in. Our system of dressing is this: soft warm woollen articles under a cloth which is impenetrable to the wind and is commonly known as box cloth; and this again under a suit of closely textured duck overalls, as snow repellers.

The feet are wrapped in squares of blanketing, and covered with leather moccasins during extreme cold, or with duck boots, having leather soles, for moderate Arctic cold or for wet.

The entire suit of clothing in wear weighs from sixteen to twenty-one pounds.

The tent furniture consists of a Macintosh floor-cloth spread upon the snow, over which is a thick duffle blanket, and upon this the men lie down in their sleeping bags, which are made of the same material, and another duffle blanket is then drawn over the party, their knapsacks serving as pillows.

It will be noticed that furs are not used. Although they are very warm and agreeable, when in good condition, to sit in, to sleep in, or even to work in, where they can be dried each night before a fire; and although they have been generally used hitherto, yet they have been deliberately set aside for such dresses as I have described because we have found that they check the escape of evaporation, they more readily absorb moisture, are more difficult to dry, and shrink much when wetted and frozen. I speak of such furs as are commonly procurable in this country. Those which have been dressed by the Esquimaux or North American Indians are much better suited to our rough work.

Let us imagine the scene when spring travelling parties set out from their ships to explore the unknown expanse before them. It was on the morning of the 4th of April that they started from the *Resolute* and *Intrepid*, commanded by the late Admiral Sir Henry Kellett and myself, at Melville Island. Out of the eighty-eight individuals composing the crews of both vessels, seventy-one were away sledging at one time, each separate sledge party consisting of one officer and six or seven men.

Each sledge hoists a gay silken banner, emblazoned with some heraldic device, some pointed motto, perhaps the name given to the sledge, or perhaps some mysterious initials known only to the leader of the small party – a little mystery, however, which only awaits the return home of the expedition for its satisfactory solution.

After mutual cheers, they part upon their lonely and toilsome mission. But trying as is the work before them, it would be difficult to overrate the enthusiasm displayed. They have just passed through many months of darkness and confinement on board, spent chiefly in preparation for this great spring effort. Nor is the keenest emulation wanting to complete a most impressive and characteristic display. Strong sense of duty and an equally strong determination to accomplish it – dauntless resolution and indomitable will, that useful compound of stubbornness and endurance which is so eminently British, and to which we islanders owe so much – certainly our colonies and our commerce, possibly even our existence as a nation.

These lonely little parties, daring and enduring so much, resemble sparks from that great fire which, I venture to say, is not yet extinct in this nation – the ardent love of the most adventurous enterprise. Each officer leads his party, selecting the route, jotting down everything noteworthy in his diary, making a running survey as they advance, and checking his estimated distances by astronomical observations. He is also constantly on the lookout for game.

When he can leave these ordinary duties he takes part in the manual labour of dragging the sledge. Clothed and fed like his men, he is housed, or rather tented, exactly as they are, sharing in all things with them; thus he becomes something more than the leader, or even the head of the party: he is its very pulse. These relations fairly established, he receives in return the most implicit confidence and devotion of his people. If he reserves anything for his own private use it is his spoon – there being, of course, no washing up of mess traps after meals in frosty weather.

In the extensive sledging operations of the third and last Government Searching Expedition, our entire immunity from severe frostbites was in strong contrast with the second expedition, where there were some thirty cases of seriously frostbitten feet, and this fact affords most satisfactory proof of the greater efficiency of the men's clothing.

Before taking leave of these spring parties, let us glance at them on the march, and notice the amount of work accomplished by those we have already alluded to.

During the month of April the snow is hard and favourable for travelling, but the winds are of course still very cold and, if at all fresh, frostbites are almost constantly playing about the men's faces. Thirst is also a good deal complained of.

May differs in being milder: the sun is now constantly up; snow blindness is more frequent than frostbites, and to avoid it as much as possible the travellers sleep by day and march by night. Some fresh snow falls, and therefore, although the sledges are lighter, the labour of dragging is scarcely diminished. Between the old frostbites, the keen winds and strong sun, all faces are badly blistered; most noses are absolutely raw, and fingertips quite callous from frequent, though slight, frostbites.

Early in June a few eider ducks, gulls and ptarmigan appear. As the month advances the snow becomes very soft. Soon the thaw bursts forth. The land is rendered impassable by innumerable streamlets, the sea ice is flooded, and the whole aspect of nature has suddenly changed.

Matters now look serious, but frostbites are things of the past, even snow blindness is less troublesome, and the abundance of water is an unspeakable relief. Those who have soap are now tempted to use it! This, however, is the season for rheumatic pains, consequent upon the daily march through ice-cold water. It is well to avoid such late travelling as this.

The travellers return with prodigious appetites; they weigh on an average twelve pounds less than when they set out; they are reduced in strength as well as in flesh, yet they can walk for hours without fatigue, their sight for distant objects is much more keen, and their powers of observation of external objects, such as traces of men or animals, etc., much sharpened by exercise. In fact, they have advanced a stage towards the condition of the North American Indian. ...

Dog-driving is so well known that but little need be said here about it. Sometimes there was a little delay at starting, the dogs not allowing themselves to be caught and harnessed. Their harness consists of a few strips of canvas, and a single trace of about twelve feet long, the leading dog having a longer trace than the rest. Once started, they are guided by the whip, which the driver should be able to use effectively with either hand. As the dogs on each flank are most exposed to its influence, there is a continual striving to get into the middle by jumping over each other's backs, so it is often necessary to halt, pull off one's mitts, and at the risk of frozen fingers disentangle the traces which have become quite plaited up together. When a dog feels the lash he usually bites his neighbour, who bites the next dog, and a general fight and howling begin. The lash is no longer of any avail and the driver is compelled to restore order with the handle of his whip. The journey is then more briskly continued for a little time, and so on throughout the march, until at length camping time arrives.

The moment our weary dogs were allowed to cease dragging, they fell asleep and remained motionless until the cook for the day commenced chopping up the pemmican or the dog's meat. At the first sound of his axe they would spring up and surround him like so many famished wolves, darting upon any splinters of meat which flew off, or watching for an opportunity to steal some pieces. Besides this severe trial of the cook's temper, more of his time was spent in chopping at the dogs than in chopping up the frozen supper. We were careful not to feed the dogs until an hour after halting; when that time arrived, their food (commonly frozen seal's or bear's flesh) was strewed over the snow, and trampled into it, before the rush for supper to enable the weak ones to secure an equal share with the strong. I think this was the only care we found it necessary to bestow upon them. We were, of course, obliged to take numberless precautions against them, removing out of their reach anything which they could eat or gnaw.

Dogs are most useful when dispatch is required, or when the temperature is so low that it is undesirable to expose more men than is absolutely necessary. Two men, with a good team of a dozen dogs, can travel with astonishing speed, the men securing themselves each night against frostbite in a small snow hut or burrow, when they can find a sufficient depth of snow to do so, but this is by no means always the case on sea-ice at a distance from the land. ... Our largest and best dogs measured twenty-three inches high at the shoulders, and weighed about seventy pounds when in fair condition. Two dogs require the same weight of food as one man, and they will draw a man's full load for about one-fourth a greater distance than a man would. If both man and dogs are but lightly laden, the dogs will almost double the distance which the man could do.

From "On Arctic Sledge-Travelling" by Admiral Sir F. Leopold McClintock, in *The Antarctic Manual for the Use of the Expedition of 1901*, edited by George Murray: Royal Geographical Society, London, 1901.

The *Discovery*, ice-bound base of Robert Falcon Scott's expedition of 1901–04.

Skinned penguins and bloody seals

Edward Wilson, known as Uncle Bill to the other members of
the expedition, was zoologist, junior surgeon and artist to the
1901–04 *Discovery* Expedition led by Robert Falcon Scott. The
Discovery spent two years frozen in ice in McMurdo Sound,
with the crew living on board or in a prefabricated hut built
on nearby Ross Island. The scientists made many significant
scientific observations and geographical discoveries during
their stay in Antarctica, but Wilson, a keen ornithologist,
seemed equally enthralled with the seabirds and marine
mammals he encountered on the voyage south.

When the expedition left the Isle of Wight in August 1901,
Wilson was twenty-nine years old and had been married
just three weeks, to Oriana (Ory) Souper. In this excerpt
from his diary, first published in 1966, the *Discovery* is in the
Southern Ocean, four days' sailing south of New Zealand.
The "Shackle" to whom Wilson refers throughout this piece
is Ernest Shackleton, the ship's third officer, who would go
on to lead his own expeditions to Antarctica.

Christmas Day 1901 – Read through the Holy Communion service before
eight o'clock, and at 10.30 we had the proper service for Christmas
Day and any number of hymns, and with our new organ it was really
very nice indeed. I opened a tin box which had been given me from
home and found sweets in immense quantities, and a present from
the parents – a box from Mother, not to be opened 'til July 1902, a
photo case, very nice indeed, some toffee from Amy, etc. etc. Very, very
good of you, all good people. We had them all shared in the mess from
day to day afterwards. Auntie's box is not going to be opened until
June 21, our real midwinter. I thought it would be more appropriate

then, and much more of a joy when we have gone through some of our exile, and half our winter darkness. I had a bundle of letters too from Hilton and a piece of holly in a letter from Con. As for Ory's present, she knows what I think of that: it is simply delightful. We had really beautiful weather, clear, fine and sunny. And we are all sunburnt already. A good deal of the day I spent alone too in my cabin. It was by no means an unhappy Christmas for me.

26 Dec – Beautiful weather, fine, brilliant sun, blue sky and no sea, very little rolling. Spent most of the day on deck trying to catch albatross, of which a great number are following us. No success at all. Everyone in great spirits and good hopes. Fair wind. Writing in the evening.

27 Dec – Brilliant weather, hot sun, but wind dropped and sails all hanging loose, almost becalmed. So we again tried for albatross and caught one, a *Diomedea melanophyrs*, which I immediately set to work on and painted in my cabin, head and feet. Writing in the evening.

28 Dec – Painting albatross all the morning. Still becalmed, no less than four albatross caught today. Set to work skinning in the afternoon and evening. We are now going six to seven knots under steam and fore and aft sail.

29 Dec – Steam dropped again as we now have an easterly wind, which, with no sea running, does us very well. Four more albatross caught today: two varieties of *melanophyrs*, one, *Thalassogeron culminatus*, a very handsome bird. Read Holy Communion 7.30 a.m. General service at 10.30, with organ and hymns most successful. Painting birds the whole evening and afternoon. With Shackle on the bridge in the evening. Beautiful skies nowadays, a splendid sunset. Last night at two a.m. just as the sun was rising, Shackle called me to see an aurora. It was a pale yellowish wreath, with shifting lights all along from east to west across the zenith, very delicate and beautiful, like a narrow strip of shimmering drapery. Now and again one saw faintly rainbow colours in it.

30 Dec – Very hard to realise what time of year we are living in. Once more the end of a month has come round, the end of the fifth month since we parted, nearly half a year gone by. Fair wind and splendid weather continues, though at midday it clouded grey all over with a grey sea and black and white porpoises, a typical outlook of the approach to the ice pack. We all eagerly look forward to entering the pack on the first day of 1902. Everyone in the best of spirits, and as keen for the fray as possible. Very few birds today, two or three albatross and a few whale birds. I spent the morning painting *culminatus* and the afternoon skinning the same. We have now six good albatross skins: five *melanophyrs* and one *culminatus*.

31 Dec – The last day of a year very full of happiness to me. Strange to think how little we knew this day a year ago of all that it has brought to us. And what a strange and novel experience is to be gone through the coming year. And yet could anything be more strange or novel than the experience of the year that has just gone by? And a whole year must pass now before a word of news can come of all that is most dear in life, and we must feed our happiness with hopes and recollections and trust God. Well, I know well that three weeks with my Ory is food enough for three years' hope, and three years' happiness. God keep her.

Today has again been grey and cold and foggy. No clouds, no sun, no sea, no ice, but a fair wind from the north. Spent the whole morning grinding and sharpening my skinning knives with Skelton, and the afternoon skinning albatross, also with Skelton. Have now eight albatross skins.

1 Jan 1902 – We all sat up to see the New Year in and sang "Old Lang Syne" over a bowl of whisky punch. Unhappily, as there were no lemons on the ship, lemon essence was used out of a small bottle by the punch maker, who, not knowing what a poisonous thing he was using, put in about an ounce. The result was that we all of us began the New Year with a splitting headache; only two or three escaped. I had very little, nothing to speak of, but two or three were very bad indeed. I spent the whole day nearly on deck. The morning was bright and sunny and I sat and read on the poop. We had six or eight albatross following us today, and several big skuas, etc. All the talk now is about sighting ice, and bets are freely knocking about, even to bottles of champagne. Most bets on this ship are small ones, a glass of port for example, or a big dinner when next we come into Lyttelton! The evening I spent painting porpoises, of which we have seen numbers lately, a very handsome little black and white porpoise.

2 Jan – Spent the whole morning on the bridge, a grey sky and a grey sea and a good deal of fog. Soon after lunch, when Royds was on watch he sighted our first ice, a fair-sized berg on the horizon, which had cleared considerably. Soon Skelton sighted another and I a third and then there were never less than four or six in sight at once. Several we passed close to and enjoyed the blue depths and the pure whiteness of the flat tops, and the caves and grottos into which the spray dashed and rose sometimes to 80 or 90 feet. It was a fine sight.

We soon picked up the ice birds, namely *Thalassoeca antarctica* and *Priocella glacialoides* and some cape pigeons and Wilson's petrel. Skelton caught three Antarctic petrels on a line in the evening.

3 Jan – We had a fine two hours out in a boat in the pack today. We woke to find ourselves in the loose pack and saw two or three single seals on the floes in the morning, a fine bright day. Then we saw two seals not very far from one another, and sail was lowered and a boat, and with Shackleton, Skelton and half a dozen men I went off after these seals. Shackles commanded the boat. We had to work our way as much by shoving as by rowing, in between the small floes, 'til at last we got near enough and Skelton got out on the ice and shot the seal. It was such hard work moving about that after hauling this seal into

our whaler we made our way back to the ship. At times we rowed, at times we shoved ourselves through lanes and in between pieces of ice, and at times we had to get out and pull, every one of us. I carried a shotgun for birds but got no chance. We got our seal on board.

After lunch set to work with Walker, our northern whaler, and skinned this seal and made his skeleton. It was a winter-coated *Lobodon carcinophagus* or crabeater – stomach full of shrimps (*Euphausia glacialis*) and remains of a fish. After tea we were sounding and dredging, and as the ship was stopped and we could see another seal on the ice, we got out a volunteer boat, Royds commanding, Barne and Skelton and I in it with six men. Barne missed the seal and it got into the water and went off of course out of sight. It is very nice getting away so in a boat. The sight is beautiful – hummocks of ice piled up in endless confusion, blue and green in the hollows and lilac distance. The dredging was not very successful but the sounding was all right. At some 3,000 fathoms bottom was found, and the wire recovered, which is very exceptional!

4 Jan – After a most sound night's sleep, before breakfast some seals were sighted and I got leave to have the ship run towards them as they were on the edge of a floe not far ahead. There were four on it, and Royds shot three of them from the bows of the ship. Walker jumped down on the floe and tied a rope round the armpits of each and they were drawn up on board. The way the blood jet flows out … is disgusting – a steady dark red curved stream a foot high or more. We then drove into two more close by and got both. Royds shot one and Skelton one. I shot and procured a beautiful snow petrel. I then had breakfast and immediately after weighed in with Walker, Weller, and Cross to skin our five new seals.

We got them skinned, measured and cut up by teatime. They were all *Lobodon carcinophagus*, four females, one old male. One of the females was in the beautiful dark chocolate summer coat.

I had an hour or two below with Shackle before dinner. After dinner we saw our first emperor penguins, two on a floe, some way off the ship, but as we had already nearly passed them we left them. They looked very large indeed for birds, much more like a small seal, the one lying down, but the other was on his hind legs. Turned in after a hard day's work. Seal skinning is all stooping and very trying. Tomorrow is to be kept as Christmas day.

5 Jan – Our Christmas Day – and a very festive one. We saw our first Adélie Land penguins today in the morning. Peace and goodwill we called them and they were not sacrificed to science, though one was caught and let go after a run on deck. We wandered along all the morning and at midday tied up to a large floe. Skis were got out for all hands and the afternoon and evenings were given up to trying them, which proved a fairly amusing exhibition – officers, staff and men staggering about in all directions, everyone thoroughly enjoying it. We saw a good many more crabeater seals – and killed two of them, both in the mottled summer coat. There were a good many snow petrels about, also Antarctic petrels, Wilson's petrel, giant petrel, and McCormick's skua. We had a most festive but sober dinner, and a sing-song in the evening.

I opened my Christmas boxes and we had Miss Lander's splendid cake in the middle of the table with a sprig of holly sent me by Constance Souper in it, Aunty's crackers all over the table and Christmas cards. One box of crackers I gave to the warrant officer's mess. Also Ory's Christmas cards I gave round to the men and they thought it was very thoughtful of her. We made souvenirs of our cards in the wardroom by getting everyone's signature to them. As it was Sunday we had morning service with Christmas and New Year hymns and I had Holy Communion in the morning. I shot a snow petrel on the ice. Late in the night just before I was turning in we saw two emperor penguins and the skipper turned the ship towards them and shot both. They proved to be young ones, which I believe have never been procured before. They were no bigger than king penguins and a very useful addition to our collection. ...

6 Jan – Spent the whole morning skinning one of the emperor penguins, the whole afternoon also, and got him finished. We killed our first Ross seal, the rarest of the southern seals, today. It proved to be a female and had a good firm set of small teeth. This was before dinner, and soon after dinner we saw two emperor penguins on a floe of ice. One was shot, the other missed, so I dropped by a line from the ship's bows on to the floe and gave chase and caught him untouched. Both these were again young emperors; not one of the four has more than a tinge of yellow on the neck and only one has a black throat. We have seen a good many Adélie penguins, both old black-throated and immature white-throated with black chins. We saw a good many crabeater seals but only one close at all, and he was right under the ship's side on an ice floe. We saw him very well. This seal has a long head and neck and a very turned-up nose, a snouty snaky-looking beast. We saw fulmar, snow, Wilson's and Antarctic petrels. Only one of the *Priocella glacialoides*.

7 Jan – At 3 a.m. this morning I was called up by Shackles to look at a seal we were to pass close by. He was lying on an ice floe ahead. When we came up to him and our bows were abreast, he was only a hundred feet off and I saw he was a sea leopard, so down I rushed to the skipper, woke him and got leave to stop the ship and go for this seal. Skelton and Shackleton shot him, though he required four or five shots and very nearly reached the floe edge. He was near a ton in weight and took the whole watch to hoist him inboard, an enormous beast with a mouth full of teeth and a head bigger than a polar bear. He was a few inches short of eleven feet.

I turned in again in an hour and a half, and an hour later was called by Royds to see three crabeater seals on a floe close to the ship. One was white, one dark and mottled and one half and half. I didn't dare call the skipper again as we had killed so many crabbers, and he hates the sight of our butcher's shop and the time it takes collecting seals. His main idea is to get through the pack. So I turned in again after seeing them. I spent the whole of that morning and afternoon painting two heads of the emperors.

Later on in the afternoon we came across three crabeater seals again on a floe and I got leave to get them as we were in thick ice and not making much way. We got the whole three – Royds and Skelton and Shackle shooting – and hoisted them all on board. They were three bulls in the white winter coat. After dinner at eleven Shackle again called me up to see a seal and it proved to be a Ross again, so we got leave and Skelton and I shot it, and then got it on board. Soon after this [I] saw a very pure white crabeater just under our bow. These seals are by no means timid. They open their mouths wide at the ship and stare with heads up, and make off a bit and then turn round again and sometimes go to sleep. They very rarely leave the floe for the water until badly wounded. They move in two definite ways, slowly by a movement like a stick caterpillar or fast by a serpentine walk, using fore flippers alternately like feet. They are very beautiful when a little alarmed and gaping at one with head erect and mouth wide open, big dark eyes and distended nostrils, and red tongue and white skin and teeth.

We had another halt to take in snow for making drinking water, and Skelton and I went after a seal for a photo but missed him. It was possible to walk miles and miles and miles over the floes today with a jump here and there from one to another. Saw a black finned whale in one of the leads.

8 Jan – Early in the morning before breakfast we had the amazing good fortune to get a third Ross seal ... This Ross seal, like the two others, was very full blubbered, and yet not one had any food in the stomach or intestines, both of which were crammed full of worms – nematodes and tapeworms. The southern edge of the pack was covered with traces and tracks of seals and penguins, evidently a happy hunting ground – brick-red stains of their excreta all over the snow on the floes. We saw here several crabber seals, and one of them had a young one with it of [blank] in length. We saw here also one immature emperor such as we have already caught, and so we left it as we were under sail, which entails the expenditure of much time and work to stop the ship, a trouble I had much compunction in giving.

Later on in the day – I had spent the whole day from breakfast till dinner time skinning and cutting up seals, bathed in blood from head to foot – we sighted the mainland of Antarctica and a more glorious sight than that midnight sun can hardly be imagined. I remained on the bridge 'til 12.30 at night with Royds and we had a glorious sort of semi-sunset effect, with immense snow peaks rolled up in a mass of golden clouds and flooded with golden sunlight. It was a sight to remember.

Strangely beautiful today has been altogether, for all through the pack ice we have not seen the sun, nothing but a dead grey sky and the glare of the ice, and the mist of the pack, often fog and thin snow, but at once on the edge smiling water and a swell and blue sky and sunshine. It was really very beautiful and may account for the extra abundance of life, even maybe the looked-for nursery of the pack ice seals, the crabeaters, who know the value of sunshine for their young. There were any number of Adélie penguins in the water in small companies.

And so we entered Ross Sea. The southern edge of the pack is very strangely defined. The swell was lost in less than a quarter of a mile, and brashy bits of ice made a welcome swish-swosh as we cleared the silence of the heavier pack and were quickly sailing south in open water.

On January 9 they arrived at Cape Adare, where a British
expedition, led by Carsten Borchgrevink, had set up base in
1899. Louis Bernacchi, a Belgium-born and Australia-educated
physicist and astronomer who was on the *Discovery* Expedition
with Wilson, had also been on Borchgrevink's expedition,
along with a number of British and Norwegian scientists and
two dog handlers from Lapland.

9 Jan – A day to be remembered, for we landed on Antarctic
continent – at Cape Adare, Camp Ridley, where Bernacchi had lived
so long with Borchgrevink. Began the day by sketching all that was
visible of a long rocky coastline of perpendicular cliff, everything above
six or eight hundred feet being covered in low fog. Well, it took us the
best part of the day to make our way along the coast, working in and
out through long stretches of drifting pack ice, very heavy humped-
up stuff, with every here and there a floating island of ice, one of the
typical southern tabular icebergs in all stages of perfection and decay
and demolition. Some were big ones a mile or two in length and a few
hundred feet up from the water. Penguins were everywhere, and seals
here and there on the ice floes, chiefly crabbers and a few Weddell, our
first introduction to the latter, on this coastal ice.

About 5 p.m. we at last rounded Cape Adare and could see, past
two rocks (the "Sisters"), the flat triangle of shingle on which stood the
huts of the Southern Cross Expedition and some millions of the Adélie
penguin. Such a sight! ... The place was the colour of anchovy paste
from the excreta of the young penguins. It simply stunk like hell and
the noise was deafening. There were a series of stinking foul stagnant
pools, full of green confervae and the rest of the plain was literally
covered with guano. And bang in the centre of this horrid place was
the camp with its two wooden huts, and a midden heap of refuse all
round, and a mountain of provision boxes, dead birds, seals, dogs,
sledging gear, ski, snowshoes, flags, poles and heaven only knows what

Huts at Cape Adare built by the 1898–1900 British Antarctic Expedition led by Norwegian Carsten Borchgrevink. When Scott's expedition visited the camp in January 1902, Edward Wilson described it as "a horrid place" with "a midden heap of refuse all round and a mountain of provision boxes".

else. Well, we anchored off the triangular flat and it was a glorious evening, as beautiful as anything could possibly be, and as quaint. It reminded me strongly of Dumont d'Urville's pictures of his Antarctic book – icebergs all round, penguins everywhere.

We landed in one of the whalers, and all went first to inspect the hut, in which were all the bunks. Bernacchi showed us all the details, and which was Borchgrevink's, Klövstad's, Hanson's, his own and the Lapps'. The latter had arranged sliding doors to their bunks which they closed quite at night, so that they actually slept in closed boxes and lived. Ten men lived in a hut, and cooked and worked and slept and ate for fifteen months, no bigger than the dining room at home and nothing like as high. We found bundles of Union Jacks lying about and posted two on the roof. I took one away with me for a souvenir. The litter round the huts was very interesting and the waste excessive. Skis, Canadian snowshoes, bamboo poles, dead dogs, seals and bundles of birds, penguins and provisions were all mixed up in a hopeless confusion and the hut looked like the centre of a rubbish heap. Yet inside the provisions were in excellent order and the place looked clean and neat enough.

There were two huts, one for living in and one for stores, and they were close together so that the intervening space, when covered over, formed a sort of antechamber. Well we were given six hours on shore and Shackle and I agreed to go in couple. He took his camera and I a gun. We started off across the penguin flat and made for the place where we had seen the white giant petrels as the ship rounded the cape. We found them there standing on the shore ice, and I killed four white ones and one dark one as they made off for the ice floating on the sea. One I lost and it floated away on the tide. I also shot two McCormick's skuas, and Shackleton two. There were any number of them flying over the penguins, ready to pounce on the first nestling found unprotected.

The penguins as usual were very amusing. They had eggs, some of them, carefully laid in a nest made of pebbles. Some had just hatched out and some had big downy or woolly young ones, more than half as big as they were themselves. These they tried to sit on, but all they could cover was the head, so it looked very absurd. Although the bird lays two eggs, hardly ever more than one young one survives. Then they had a strange way of flying at one very often. A big penguin would start running towards one from some considerable distance, and without a moment's hesitation would seize one's trousers and begin battering one's shins with its very bony flippers. The number of these birds was simply incredible. They covered the 200 odd acres, they covered the sides of the mountain [1,000] feet high and they spread in small colonies over the top. There was a constant stream of them up and down the side of the mountain and their paths were very useful to us in getting up the steep snow and rocky slopes.

We found several skuas on the way up, with their fledglings, pretty little greyish-white fluffy things with pale blue bill and feet. But the old birds kept up a continual attack on us when we were anywhere near their young, chattering excitedly and dashing at our heads so boldly and persistently that we had to protect ourselves with a stick and one of the birds was killed by a hit on the head. We worked our way to the top and began to hunt for Nicolai Hanson's grave, which was marked by a big stone and an iron cross. But nowhere could we find it. Somehow we missed it, for we went a mile and a half over the top hunting for it. I was very anxious to see it. …

A snowstorm came over the mountain while we were on the top so we gave up the search and came down the rocky sides. On our way down we saw several Wilson's petrels flying about the face of the rocks, so we watched them and soon marked down two crevices into which the birds went, but neither were within reach without great risk, so we watched again and got a third one in a bank of snow under a big boulder. I couldn't get my arm in so we began chipping the ice away from under the stone. It took us nearly three-quarters of an hour. Happily I had a sheath knife and we used a stone for a hammer. At last I could reach the end, the whole length of my arm – it was up to the shoulder – and out of this burrow we got one bird, which I had shot flying about outside, but also two more adults inside, a dead one also, and two eggs, one fresh and one addled, and a large handful of penguins' feathers forming the nest. In firing at the little petrel, although I killed it dead I also killed five penguins, which gives an idea of how thick they are on the ground for I only used one barrel.

Edward Wilson sketching on the Beardmore Glacier in 1912, during his second and fatal expedition to Antarctica with Scott.

After this we made our way back to the ship – or rather to Borchgrevink's hut, where several photos were taken and a quantity of rubbish collected, which each man thought might come in useful, from the rubbish heap – and then back to the ship.

Skelton had taken some excellent photos. Armitage, Bernacchi, and Barne had spent all their time taking magnetic observations. Everyone was thoroughly satisfied with an excellent day on shore. Bernacchi was so anxious to see that Hanson's grave was all right that after dinner he started up and found it all safe, and with Barne and Koettlitz brought home a number of skua fledglings and young penguins in different stages and many of their eggs. Koettlitz was kind enough to blow all these eggs for me, a true labour of love, as nearly every one was rotten.

It was a glorious sight from the top of Cape Adare in the afternoon, with a hundred icebergs at the mouth of the bay and pack ice in big stretches, and bright sunshine and the red acreage below us teeming with penguins, and the old black ship just off the shore reflected in the calm clear water. And again at night with heavy golden clouds all over the snow peaks and the glaciers of Robertson Bay. It was glorious, a flood of golden light in the midnight sun. At three a.m. we were off and our visit to Cape Adare was over. We left letters in a cylinder in the hut for the relief ship to pick up.

A month later the *Discovery* entered McMurdo Sound, where their two years of scientific work began. Wilson would be chief of scientific staff on Scott's next venture to Antarctica, the *Terra Nova* Expedition of 1910–13. In the winter of 1911 he led the expedition that Apsley Cherry-Garrard later wrote about in *The Worst Journey in the World*. In 1912, as Scott's team returned from the South Pole, Wilson died with Scott in a tent on the Ross Ice Shelf.

From *Diary of the Discovery Expedition to the Antarctic Regions, 1901–04* by Edward Wilson: Blandford, London, 1966.

German scientist Erich von Drygalski led the German South Polar
Expedition of 1901–03 on the *Gauss*, which became trapped in ice about
fifty kilometres from land. This picture shows the hydrogen-filled balloon
from which Drygalski measured atmospheric temperature and humidity
and admired the view.

Drygalski's balloon ascent

Erich von Drygalski, a geography professor from Berlin, led the first German expedition to Antarctica in 1901–03. As with many early Antarctic expeditions, the sea ice was a problem and their ship, the *Gauss*, spent fourteen months trapped in the Southern Ocean south of Australia, some fifty kilometres from the Antarctic coast.

The German South Polar expedition made many scientific discoveries and innovations, including recording penguins' calls on a phonograph and spreading dark material on the sea ice in order to melt out a path through which the trapped ship could escape. On his return Drygalski published twenty volumes of scientific findings and two atlases.

This passage recounts the events of a day in March 1902. One of the significant landmarks in the area – which Drygalski refers to in this passage – was a 370-metre-high volcanic cone, which they named Gaussberg after the ship. The psychrometers Drygalski has trouble with on the balloon ascent were instruments containing wet and dry bulb thermometers designed to measure air temperature and humidity.

March 29th (Saturday before Easter) was a fine day, probably the best we had, and one of the few on which a balloon ascent in the Antarctic was even worth considering, so that things turned out very fortunately in this respect. Sixty-three gas cylinders were used to fill the balloon. Of these, three were found empty to begin with and some others were not completely full, but such losses had already been discounted in advance. This was in any case quite enough to inflate the balloon well beyond what was necessary, for while I was aloft it became so distended because of the intense heat of the sun that I had to open the vent several times in order to prevent the envelope getting too full

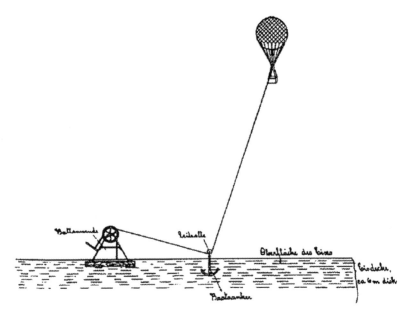

This sketch showing how the balloon was secured to the ground was drawn by the expedition's chief engineer, Albert Stehr.

or bursting. This job was itself quite difficult because the ventilator springs had been set too stiffly, and I realised that if, for instance, the anchor cable should part, speedy deflation by means of the ventilator would be out of the question. In any case the balloon would be carried a considerable distance from the ship by the wind before it could be brought down.

The entire ship's company was involved in the ascent under Stehr's guidance. Twelve men held the balloon, two manned the gas cylinders, opening the taps and, as this had to be done in two groups, effecting the necessary exchange of cylinders. The upward force was very strong so that the first time the balloon was released for testing without the basket being manned it proved impossible to bring it down with the winch, and it had to be hauled in by hand from a height of 100 metres.

Even when I was aboard it went up quickly in the almost dry air, which only at some height gave way to occasional light puffs of wind, and these died away again completely above 300 metres.

The movement of the air caused the balloon to twist about slightly, which made taking bearings difficult. In the ascent I received a number of signals via the telephone, mostly warning me to open the ventilator because the balloon was inflated too far. I had with me an aspiration psychrometer and a sling psychrometer to deal with, and enough to do apart from that to get bearings. Unfortunately, the cord on the sling psychrometer parted, so that the instrument fell from a height of 100 metres. The aspiration psychrometer also stopped because of an accident, so that the exact temperature measurements at height were not completed. I did observe, though, that it became warmer as I ascended. At 500 metres it was so warm that I took off my gloves and chose a light cap without earflaps, which I was also able to do without when I accidentally dropped it over the side at the same height.

The radiation from the sun was extraordinarily strong, but the reflection from the ice surface below had little effect at height and snow goggles were unnecessary. Up to a height of 100 metres I could hear everything that was said below, even quietly, and louder shouts were audible even higher; I had understood most of the messages before they were entrusted to the telephone. The surrounding view from 500 metres was grandiose. Above about fifty metres the newly discovered Gaussberg came into view, and from a greater height it became evident that it was the only ice-free point for a long way around.

The inland ice rose prominently to the east, evidently a high mountain region but apparently entirely ice-covered. Very likely it was the high land we had glimpsed on the morning of the day before becoming beset. Immediately in front of this was the main assemblage of icebergs, and the great colossi which surrounded us shone in the reflected light. Further to the west, too, there were icebergs at the land's edge, often so placed that they were embedded in the surface of the sea ice on one side and the inland ice on the other. For this reason it was sometimes hard to tell where the edge of the inland ice was, especially in the west.

It was, however, possible to see that the coast took an east–west line for the most part, with shallow indentations to the south, westwards of the Gaussberg. The smooth surfaces of the sea ice seemed to rise gradually towards the land, so that it was only by virtue of the northward facing edges of the icebergs that the lower level of the sea ice was in evidence, thus giving an indication of the edge of the inland ice that faced towards us. It was noteworthy that the open cracks in the floe field all around us were in every case aligned towards S20°W (magnetic) – that is, more or less SE, like the lines of icebergs descending from the high land.

Not far to the north of us the open pack ice began, in which the leads and cracks were wider. South of the *Gauss* was the smooth ice field traversed by the sledge party. Among the floes to the north were many leads, gleaming in the sunshine, but nowhere was there any open sea visible any more. The most common form of iceberg was tabular, but there were also those with low rounded tops in fairly large numbers, of the kind we had seen earlier in the voyage, especially in the south. The inland ice behind the Gaussberg seemed to rise towards the south, quickly at first and then more slowly, until the view vanished in the high, far-off distance; to the east of the Gaussberg the surface of the inland ice exhibited all sorts of variations near the edge. I could recognise the point opposite which we had first been on February 21st. The summit of the inland ice we had seen there was smooth, and marked only by the pattern of crevasses that spread across it. Very obvious, too, was a depression in the inland ice running from east to west, perhaps turning to the south far in the western distance.

From up above it was clear to me that when our floe ice started to break up our way would lie to the west; in that direction it might be possible to penetrate a little further south with the ship. The Gaussberg was the only notable interruption and the only anomaly in the otherwise infinite monotony. It stood out clearly at the edge of the inland ice surrounded by deep melt hollows, an isolated rounded dome covered in snow on its western side facing the east winds. Behind it the ice rose to greater heights. Winding the balloon down after I had spent almost two hours in the airy heights was as easy a task as raising it; the only stops came as each hundred-metre length of its anchor chain was unshackled in turn.

From *The Southern Ice-Continent: The German South Polar Expedition aboard the Gauss 1901–1903* by Erich von Drygalski, translated by M.M. Raraty: Bluntisham Books / Erskine Press, 1989.

Otto Nordenskjöld in his Swedish Antarctic Expedition's hut on Snow Hill Island off the east coast of the Antarctic Peninsula, 1902. The expedition used the hut as a base for scientific exploration of the coast and nearby islands.

Luxuriant vegetation and extensive coasts

Otto Nordenskjöld, a geologist at Uppsala University, was the leader of the 1901–04 Swedish Antarctic Expedition. The expedition's ship *Antarctic* left Sweden in late 1901 bound for the Antarctic Peninsula. In February 1902 Nordenskjöld and five other men were dropped off at Snow Hill Island, latitude 64°28', near the northern extent of the peninsula, and the ship returned to Falkland Islands for the winter.

The six men set up camp on a narrow cape that Nordenskjöld described as having a low shore "that seems to us to have been created for the very purpose of establishing a wintering station". The expedition had a full scientific programme and made contributions to geology, meteorology, hydrology, cartography, zoology, botany and microbiology, but the highlight was fossil discovery. In 1893, on an earlier expedition, the ship's captain Carl Larsen had made the first Antarctic fossil finds at nearby Seymour Island but the new finds were significant: they suggested that fifteen million years earlier this part of Antarctica had had a warm temperate climate.

This extract from Nordenskjöld's account of his stay in Antarctica starts in December 1902, when he and three companions set off by sledge to Seymour Island. While Nordenskjöld explored the geology at the southern end of this island, the rest of his team travelled on north to Cockburn Island. The fossils brought back from Seymour Island included some uncommonly well preserved ammonites.

I started on the 2nd December, accompanied by Jonassen and Åkerlund, and reached the depot without any great difficulty. Farther out on the ice we could see large crowds of seals, amounting to several hundred in number.

I at once made for the newly discovered cross-valley, and in a grey shale among the shore rocks near its southern entrance I caught sight, for the first time, of something which made me surmise that petrified wood was not the only vegetable fossil of Antarctic regions. I went along the slope of the valley right across the island, studying and collecting from each knoll, but although I everywhere found traces of vegetable petrifications I could not succeed in finding one which allowed of any determination. I returned to the camp late in the evening, and enjoyed a good supper of fresh penguin meat.

Some of the penguins which had for the second time been deprived of their eggs had gone out to sea, but otherwise they did not seem to make much to-do about the thefts. A few lay quietly in the empty nests as if nothing had happened, while others were said to have carried rotten or cracked eggs to their rifled homes and sat upon them.

The next morning, December 3rd, I went out early along the shore, past the cross-valley and towards the north headland of the island, which here forms a high, level, extensive plateau. I did not ascend the plateau on this occasion but stayed on a little terrace some distance below the top, which was traversed by valleys and had small irregular knolls of hard rock. In this place I made a most interesting discovery consisting of the loose scattered fossil bones of a vertebrate, but I could not investigate the matter more closely on the spot.

A more important discovery of its kind could scarcely be made by such an expedition as ours. One of the greatest problems in the investigation of the geography of ancient epochs concerns the role played by the south polar regions during that important period when higher animals and plants of modern types began to appear upon the Earth – i.e., during the latter part of the Cretaceous and the first part of the Tertiary systems. The distribution of land organisms on the southern half of the globe presents many peculiarities, especially in the circumstance that nearly related forms occur in South America, Africa and Australia, while these continents are not now in any way connected by land. These phenomena could best be explained could we suppose the existence of a great mass of land around the South Pole, formerly continuous with these three divisions of the globe, and across which animals and plants could migrate from the one side of the Earth to the other. But no proof of the correctness of such an hypothesis had been discovered before the advent of our expedition to Antarctic regions.

The discovery just mentioned has not, however, been able to fully decide the correctness of the theory. Apart from some large and as yet undetermined bones, nearly all of those that I found belong to a species of penguin considerably larger than the largest now living – the emperor penguin. It is true that this discovery is in itself of exceeding interest as it demonstrates that even at such a distant epoch – probably the beginning of the Tertiary period – the penguin was an inhabitant of the Antarctic regions. But for the solving of the general problem, it is clear that it would be of still greater importance to discover here the remains of land vertebrates.

I was to make another discovery on this memorable day, which strongly supports the hypothesis mentioned above. When I came back to the cross-valley I stopped there to continue my search among its rocks for plant fossils. I looked a long time without finding anything but fragments, until my eyes fell upon a brown, coarse, hard, tuff-like rock, and in this I at last found what I had sought for so long: numerous large and quite distinct leaves – although, as a rule, by no means well preserved or easily determinable – belonging to a variety of different forms of exogenous trees, firs and ferns. It is difficult to express the joy I felt at this moment. Could it have been a dream which led me to choose just these tracts for my field of labour? For if there was one hope whose fulfilment or non-fulfilment was, in my thoughts, almost

Antarctica was once a verdant land, and the eight scientists on the four-year Swedish expedition collected numerous fossils dating back to the Jurassic period. This *Cladophlebis* fern fossil was found at Hope Bay on the peninsula.

synonymous with the success or failure of the expedition, it was just that of being able to discover in these regions determinable Tertiary vegetable fossils.

Our investigations have proved that these islands are built up of a connected series of deposits, which become more recent the more one comes to the north. The oldest strata, which are found in the district immediately surrounding the station, belong to the middle or upper Cretaceous system, and contain numerous ammonites and mollusca, as well as sea urchins and crustaceans. On Snow Hill Island the fossils are, in general, not very well preserved; the opposite is usually the case on Seymour Island, where, moreover, the store of fossils is greater. Ammonites are found on the last named island too but belonging to other types, and it seems most probable that the deposits there belong to the youngest chalk formation. Ammonites are wanting, however, in the northern part of the island, their place being taken by numerous new forms of mollusca, brachiopods, *Encrinus liliiformis*, etc., and it was in these deposits, too, that the fossil bones and leaves found were discovered.

The collections of fossils we have brought home will be the thread which will gradually lead to discoveries enabling us to form a picture of the chief features of the nature of the Antarctic regions, from the Jurassic period down to our own times. And it must be remembered that it is a continent which has thus been opened to scientific investigation, and a continent which, during the period of the Earth's development just named, was not an icy waste but a land with luxuriant vegetation and extensive coasts, where maybe many types of animals and plants were first developed that afterwards found their way as far as to northern lands.

From *Antarctica: Or Two Years Amongst the Ice of the South Pole* by Otto Nordenskjöld and Johan Gunnar Andersson: C. Hurst & Company, London, 1977.

Toralf Grunden, Johan Gunnar Andersson and Samuel Duse were put ashore at Hope Bay by the *Antarctic* on December 29, 1902 and set off on a 200-mile journey to join Nordenskjöld's team at Snow Hill Island. Forced to turn back they spent eight terrible months at Hope Bay before trying again. When they finally reached the men at Snow Hill they were black with soot from their cooking fires at Hope Bay and sported long scraggy beards.

Penguin eggs and fried seal

The Swedish Antarctic Expedition's ship *Antarctic* arrived at the Antarctic Peninsula in December 1902 and spent a week mapping the channel that separated the peninsula from the line of islands to the north-west. Geologist Johan Gunnar Andersson and his fellow scientists used every landing as an opportunity for collecting geological and botanical specimens, while a trawl was used to collect fauna from the ocean. Andersson was also interested in what he described as a "catch" of practical importance.

On several small islets we found rather large penguin colonies belonging to but one species, *Pygoscelis antarctica*, where the egg-laying season was drawing to a close, and we seized every opportunity of enriching our supply of provisions with the dainty food. Once we took a whole boat's load on board, and several casks were filled with eggs packed in salt.

The egg of the penguin is about the size of a goose egg. On boiling, the white coagulates to an almost glass-clear mass (not of a porcelain white like that of a hen's egg), a circumstance which gave rise to some hesitation in the case of certain prejudiced individuals when the eggs came on the table. But the egg of the penguin is very palatable when boiled, and does not possess that peculiar taste which renders eider eggs somewhat unpleasant. The rich supply of fresh eggs reacted upon the whole of our meals. The cook and the steward felt continually called upon to produce fresh dishes, and almost every day had its surprise in the form of a new omelette or a fine cake.

Future Antarctic expeditions, and especially those that mean to winter in these regions, cannot be too strongly advised to find out a penguin colony in good time and provide themselves with a plentiful supply of eggs.

The ship's main goal was to collect the wintering-over party – Otto Nordenskjöld and his five companions – who had been dropped at Snow Hill Island the previous summer. Sea ice prevented the ship landing at the island, so on December 29 Andersson and two other men – Lieutenant Duse and Seaman Grunden – were dropped near the tip of the Antarctic Peninsula. From here they planned to sledge to Snow Hill Island and return with Otto Nordenskjöld and the wintering-over party. Ironically, though, they encountered open water, which made sledging impossible, and Andersson, Duse and Grunden returned to Hope Bay to await the *Antarctic*.

At first they used their time to add to their already rich collection of plant fossils, finding evidence of fern, cycad and pine dating back to the Jurassic period. However, by February there was still no sign of the ship. The possibility of having to spend the winter where they were "grew gradually to a threatening certainty".

"We should soon stand face to face with the polar winter," Andersson wrote, "provided only with two storm-torn tents and an insufficient supply of food; we must in some way wrest from nature the simplest means of preserving life – shelter, food and firing." The food they had unloaded from the *Antarctic* – bread, tinned meat, barley, butter, cocoa, coffee and sugar – would do little more than supplement what they would need to hunt or catch.

They built a stone-walled hut and used the sledge, tarpaulins and tents to make a roof, a floor and a second layer of wall. In the final preparations, Duse laid a carpet of penguin skins under the floor tarpaulin to insulate against the cold ground, and Andersson "took the wing of a giant storm petrel and swept the house clear of snow".

Until the close of February, being in expectation of the speedy return of the *Antarctic*, we had lived chiefly on the provisions stored in the depot on the occasion of our landing. But at the beginning of March we made a complete alteration in our manner of living, changing hastily from enjoying perfectly civilised fare to supporting ourselves almost exclusively on the products of the land around us. ...

At the beginning of winter we had had an opportunity of laying the animal world around us under tribute – a tribute which gave us both firing and food.

On 19th of February we made our great slaughter of the penguins. The Adélie young ones were now almost ready to go out into the sea, and it was among them we chose our first victims. It was raw and bloody work but "necessity has no law." We killed that day no less than 150 penguins.

But the great mass of the penguins had already gone out to sea, and those who remained grew shyer after every hunt. After the 7th of April there were only a few hundreds of birds left on shore, all of them very shy, and we needed quite a hundred of them ere we could feel sure we had a sufficient supply for the winter. An idea came to my head that we could possibly use the loose snow, which hindered our progress so much and facilitated that of the penguins in an equal degree, to make a trap for the birds.

The plan succeeded beyond all expectation, and on 8th of April we obtained no less than 101 penguins in this way. Including those killed now and then by the cook for the day ("I'm going out to kill the dinner," he used to say sometimes), no less than 700 penguins were dispatched by us at Hope Bay. It was faithless and ungrateful of us to thus destroy the peace of an hitherto untroubled world of birds, but no one can rightly blame us for killing the number of animals we considered absolutely necessary to supply our wants – a number which would, in reality, not have saved us from the touch of famine had we not later on increased our food supplies with a number of well needed winter seals.

As the winter came on we began to think of new ways of preparing our penguin-meat. Penguin-beef fried in the fat found beneath the skin and in the entrails of the birds proved an excellent dish, and Duse made an admirable discovery – grilled penguin – the meat being rolled in the fine crumbs obtained from ship's biscuits. And one day, shortly after we had moved into the stone hut, Grunden gave us some seal meat of an extremely clean and agreeable taste. It had been fried in seal oil. (It should be mentioned that our frying-pan consisted of a flat preserve tin, with a nail and a bit of wood for a handle.)

Even while we were in the tent we began the experiment of trying to make fires with blubber, but it was first after several weeks' residence in the hut that we succeeded in overcoming all difficulties in the arrangement and care of the train-oil lamps, of which we needed two kinds, a small one for the purposes of illumination and a larger one for the kitchen. For the tent lamp we had a flat herring tin, which was filled with small squares of blubber, in the midst of which was put a wick made of tent rope. The two big preserve-tin "smoker-stoves" we had in the kitchen required such long and thick ropes, however, that it soon seemed as though we should soon have no wick material left, but we at last made the fortunate discovery that they could burn without any wick at all.

At first we had much trouble in getting the lamps to burn so it sometimes took five or six hours to boil the penguin soup, but by the end of the winter we had become real virtuosos in the art of turning the blubber into a burning mass of flames, smoke and soot, and all within the space of a minute or two.

We called our lamps by the abusive name of "smokers" and not without good reason, for sometimes when the snowstorms stopped up a chimney we had made of old tins, and which led into the open air, the smoke became so dense that we could scarcely distinguish each other's faces. The tent lamp, which burned all night in order to warm the air a little, was placed on a fireproof place in the bottom of a large unused cooking apparatus. Sometimes it burned calmly the whole night through, but as a rule it went out towards the morning. Once it went wrong altogether, the whole mass of partly burned blubber suddenly taking fire and developing a heavy smoke, which might have suffocated the whole party had not one of us awakened and put the thing out.

From the day we moved into the hut it was agreed that we should share the work equally. Every third day came each one's turn to sit on the vegetable box and prepare the plain food. When the toilsome time in the kitchen was ended, the cook crept into his sack with a pleasant lazy thought of the two free days to come. But inactivity was no lasting joy. On the second day one lay wishing to be at work again out in the kitchen, so our plan evidently provided us with a most necessary change amidst the monotony of winter life.

The *Antarctic* was crushed by ice and sank in February 1903, with the loss of most of the expedition's scientific collections. Captain Larsen and his crew travelled by lifeboat to Paulet Island, a speck of land off the top of the Antarctic Peninsula, where they set up camp for another winter. In September Andersson and his companions left Hope Bay and made their way to Snow Hill Island, meeting Nordenskjöld's party along the way. The survivors of the expedition – the Nordenskjöld and Andersson teams from Snow Hill Island and the crew of the *Antarctic* from Paulet Island – were rescued by an Argentine naval vessel in November that year.

From *Antarctica: Or Two Years Amongst the Ice of the South Pole* by Otto Nordenskjöld and Johan Gunnar Andersson: C. Hurst & Company, London, 1977.

Along the journey to the south magnetic pole, members of Shackleton's 1907–09 party took regular observations using a dip circle, an instrument that responds to magnetism in the vertical plane. The closer they got to the south magnetic pole, the closer the dip needle was to vertical. This photograph shows use of the dip circle on Mawson's later 1911–14 expedition.

Hunting the south magnetic pole

The *Nimrod* Expedition of 1907–09 was led by British Navy lieutenant Ernest Shackleton, who had first travelled to Antarctica as third officer on the *Discovery* Expedition. Two of its most significant achievements were the discovery of the Beardmore Glacier approach to the Polar Plateau, and the apparent locating of the south magnetic pole by Australian geologists Edgeworth David and Douglas Mawson and Scottish doctor Alistair Mackay.

In his diary Mawson complained constantly about forty-nine-year-old David, describing "the Prof" as "dreadfully slow ... he does nothing ... never, or seldom helps pack a sledge". In this account by David the professor seems oblivious to his former student's frustrations. When it begins, the three men, man-hauling two sledges, have been travelling for nearly three months from their base at Cape Royds and are more than 1,500 metres above sea level on the Polar Plateau. The closer they get to the south magnetic pole the less reliable their magnetic instruments become. As well as a theodolite and compass, which they used to find their way across the sea ice to Victoria Land and up on to the plateau, Mawson is carrying a dip circle: an instrument that responds to magnetism in the vertical plane. Lines of force of Earth's magnetic field loop the planet rather than running parallel to the ground, so at the south magnetic pole the horizontal direction of the compass (also referred to here as a transit instrument) would be north in every direction and the inclination of the dip needle would be vertical.

December 31 passed off without any special event, other than that after Mackay had repaired the tent in the morning it became torn again at lunchtime when we were fitting it over the tent poles. Mawson took a fresh set of magnetic observations. We camped for this purpose at the bottom of a wide undulation in the névé surface. We were disappointed at his announcement that … the magnetic pole was further inland than had been originally estimated.

What with the observations with the Lloyd-Creak dip circle and the time occupied in repairing the rents in the tent, we ran ourselves somewhat short of time for our sledging that day, and did not camp until a little before midnight. We were still dragging the sledge on an upgrade; the surface was softer and more powdery than before, and the sastrugi heavier. Also since the previous Tuesday we had been obliged to put ourselves on somewhat shorter rations than before as we had to take one-eighth of our rations out in order to form an emergency food supply, in the event of our journey to and from the magnetic pole proving longer than we originally anticipated.

That night, about a mile before reaching camp, we sighted to the west of us, much to our surprise, some distinct ice falls. This showed us that the snow desert over which we were travelling had still some kind of creeping movement in it. A skua gull came to visit us this New Year's Eve. He had been following us up for some time in the distance, mistaking us perhaps for seals crawling inland to die, as is not infrequently the habit of these animals. We were now about eighty miles inland from the nearest open water. Being disappointed of his high hopes, he left us after that day and we saw him again no more. The run for the day was about ten miles. We felt very much exhausted when we turned into our sleeping bag that night.

January 1, 1909 (New Year's Day) was a beautiful calm day with a very light gentle plateau wind, with fairly high temperature. The sky was festooned in the direction of Mount Nansen with delicate wispy cirrus clouds converging in a north-east direction. Later on, towards the evening, it was evident that these cirrus clouds were strongly bent round from south-west in a northerly direction. Possibly this bending with the concave side to the west-north-west was due to the pressure at a high level of the anti-trade-wind blowing towards the east-south-east. Mawson took observations for latitude and for magnetic deviation at noon. He made our latitude at noon to be 74°18'. That night Mawson gave us a grand hoosh and a rich pot of cocoa in celebration of New Year's Day. We all thoroughly enjoyed this meal after our exhausting march.

On **January 2** we noticed that the sastrugi were gradually swinging round into a direction a little north of west. The snow was frequently soft in large patches, which made sledging very heavy. We ascended altogether about 290 feet but we crossed a large number of broad undulations, the troughs of which were from thirty to forty feet below their crests. These undulations considerably increased the work of sledging, and the loose patches of snow were so very soft and powdery that the runners of our sledge sunk deeply into them, so that it was only with our utmost efforts that we were able that day to finish our usual ten miles.

Again we were much exhausted when the time came for camping. We were beginning to suffer, too, from hunger, and would have liked more to drink if we could have afforded it. We talked of what we would have drunk if we had had the chance. Mackay said he would have liked to drink a gallon of buttermilk straight off, Mawson would have preferred a big basin of cream, while I would have chosen several pots of the best coffee with plenty of hot milk.

We were still climbing on **January 3**, having ascended another 500 feet. It proved the heaviest day's sledging since we reached the plateau. The snow was still softer than on the previous day and the surface was more undulating than ever, the troughs of the undulations being about fifty feet below the crests. The sastrugi themselves were from two to three feet in height. The crests of the large undulations were usually formed of hard snow, the strong winds having blown any loose material off them. This loose material had accumulated to some depth in the troughs and hence made the wide patches of soft snow, which made our sledge drag so heavily as we crossed them. By dint of great efforts we managed to finish our ten miles for that day.

The next day, **January 4**, we were pleased to find that there was less upgrade than on the previous day. We were now at an altitude of over 6,000 feet and found respiration in the cold rarified air distinctly trying. It was not that we suffered definitely from mountain sickness, but we felt weaker than usual as the result, no doubt, of the altitude combined with the cold. Towards evening large clouds developed, much like the whaleback clouds which we had often observed forming over Erebus about the time of blizzards. Great rolls of cumulus spread rapidly from the north-west towards the south-east and we feared that a blizzard was impending. On the whole the sledging was a little easier today than the preceding day, and again we managed to do our ten miles.

On the morning of **January 5** we found the sky thickly overcast, except to the south and the south-east, where clear strips of blue were showing. We thought snow was coming. The weather was perfectly calm, comparatively warm, but the light dull. We could still see the new inland mountain and Mount New Zealand distinctly. The sun was so oppressively hot when it peeped out from behind the clouds that one could feel it burning the skin on one's hands.

The surface was more marked by sastrugi than ever, but on the whole firm. We sledged ten miles. I will quote from my diary the notes regarding some succeeding days.

January 6 – Today the weather was gloriously fine. Bright warm sunshine with a crisp, cold air in the early morning and the weather almost calm. The pulling was rather heavy during the afternoon; possibly the hot sun may have somewhat softened the surface of the snow. This morning I left off my crampons and put on a new pair of finnesko. These latter proved somewhat slippery, and in falling heavily this afternoon over one of the sastrugi I slightly strained some muscles on the inner side of my left leg, just below the knee. This gave me a considerable amount of pain for the rest of the journey. Mackay lost all his stockings and socks off the bamboo pole of the sledge, but was fortunate enough to recover them after walking back over a mile on our tracks.

January 7 – We were up at 5 a.m., when the temperature was minus 13° Fahrenheit. We were anxious to arrive at the end of our first five miles in good time for Mawson to get a meridian altitude, and take theodolite angles to the new mountain and Mount New Zealand, which were now almost disappearing from view below the horizon. Mawson made our latitude today 73°43'. This was one of the coldest days we had as yet experienced on the plateau, the wind blowing from west by north. We all felt the pulling very much today, possibly because it was still slightly uphill, and probably partly on account of mountain lassitude. The distance travelled was ten miles.

Friday, January 8 – Today, also, was bitterly cold. The wind blew very fresh for some little time before noon from a direction about west by north, raising much low drift. Our hands were frostbitten several times when packing up the sledge. The cold blizzard continued for the whole day. At lunchtime we had great difficulty in getting up the tent, which became again seriously torn in the process. Our beards were frozen to our Burberry helmets and balaclavas, and we had to tear away our hair by the roots in order to get them off. We continued travelling in the blizzard after lunch. Mawson's right cheek was frostbitten, and also the tip of my nose. The wind was blowing all the time at an angle of about 45° on the port bow of our sledge. We just managed to do our ten miles and were very thankful when the time came for camping.

The following day, January 9, a very cold plateau wind was still blowing, the horizon being hazy with low drift. We were now completely out of sight of any mountain ranges, and were toiling up and down among the huge billows of a snow sea. The silence and solitude were most impressive. About 10.30 a.m. a well-marked parhelion, or mock sun, due to floating ice crystals in the air, made its appearance. It had the form of a wide halo with two mock suns at either extremity of the equator of the halo parallel to the horizon and passing through the real sun. Mawson was able to make his magnetic deviation observation with more comfort, as towards noon the wind slackened and the day became gloriously bright and clear. In the afternoon it fell calm.

We were feeling the pinch of hunger somewhat, and as usual our talk under these circumstances turned chiefly on restaurants, and the wonderfully elaborate dinners we would have when we returned to civilisation. Again we accomplished our ten miles, and were now at an altitude of over 7,000 ft.

January 10 was also a lovely day, warm and clear; the snow surface was good and we travelled quickly. There was a strong "Noah's Ark" structure in the high-level cirrus clouds, there being a strong radiant point respectively in the north-west and south-east, and this made us somewhat apprehensive that we were in for another blizzard. These cirrus clouds were also strongly curved, with the concave side of the curve facing the north-east. We thought this curve was, perhaps, due to the anti-trade-wind bending round in a direction following that of the curve in the wisps of cirrus.

January 11 – We were up about seven a.m., the temperature at that time being minus 12° Fahrenheit. It was a cold day today, and we had a light wind nearly southerly. At first it blew from between south and south-south-east; this gradually freshened at lunchtime and veered towards the west. It then returned again more towards the south-south-east. Mawson had a touch of snow blindness in his right eye. Both he and Mackay suffered much through the skin of their lips peeling off, leaving the raw flesh exposed. Mawson, particularly, experienced great difficulty every morning in getting his mouth opened, as his lips were firmly glued together by congealed blood.

That day we did eleven miles, the surface being fairly firm, and there being no appreciable general upgrade now but only long-ridged undulations, with sastrugi. We noticed that these sastrugi had now changed direction, and instead of trending from nearly west, or north of west, eastwards, now came more from the south-east directed towards the north-west. This warned us that we might anticipate possibly strong headwinds on our return journey, as our course at the time was being directed almost north-west, following from time to time the exact bearing of the horizontal magnetic compass. The compass was now very sluggish; in fact the theodolite compass would scarcely work at all. This pleased us a good deal, and at first we all wished more power to it, then amended the sentiment and wished less power to it. The sky was clear, and Mawson got good magnetic meridian observations by means of his very delicately balanced horizontal moving needle in his Brunton transit instrument.

January 12 – The sky today was overcast, the night having been calm and cloudy. A few snowflakes and fine ice crystals were falling. The sun was very hot and it somewhat softened the snow surface, thereby increasing of course the difficulty of sledging. We sledged today ten and three-quarter miles.

That evening after hoosh, Mawson, on carefully analysing the results set forth in the advance copy of the "Discovery Expedition Magnetic Report", decided that, although the matter was not expressly so stated, the magnetic pole, instead of moving easterly as it had done in the interval between Sabine's observations in 1841 and the time of the *Discovery* Expedition in 1902, was likely now to be travelling somewhat to the north-west. The results of dip readings taken at intervals earlier in the journey also agreed with this decision. It would be necessary therefore to travel further in that direction than we had anticipated in order to reach our goal. This was extremely disquieting news for all of us as we had come almost to the limit of our provisions, after making allowance for enough to take us back on short rations to the coast. In spite of the anxiety of the situation, extreme weariness after sledging enabled us to catch some sleep.

The following morning, **January 13**, we were up about six a.m. A light snow was falling and fine ice crystals made the sky hazy. There was a light wind blowing from about south-south-east. About eight a.m. the sun peeped through with promise of a fine day. We had had much discussion during and after breakfast as to our future movements. The change in the position of the pole necessitated of course a change in our plans. Mawson carefully reviewed his observations as to the position of the magnetic pole, and decided that in order to reach it we would need to travel for another four days. The horizontally moving needle had now almost ceased to work.

We decided to go on for another four days and started our sledging. It was a cold day with a light wind, the temperature at about 10.30 a.m. being minus 6° Fahrenheit. At noon Mawson took a magnetic reading with the Lloyd-Creak dip circle, which was now fifty minutes off the vertical – that is, 89°10'. At noon the latitude was just about 73° South. The sastrugi were now longer and higher than usual and there were two distinct sets. The strongest sastrugi trended from south to north, a subordinate set from south-east to north-west. That day we sledged thirteen miles.

January 14 – The day was gloriously clear and bright with a warm sun. A gentle wind was blowing from about south-south-east, and there was a little cumulus cloud far ahead of us over the horizon. The surface of the snow over which we were sledging was sparkling with large reconstructed ice crystals, about half an inch in width and one-sixteenth of an inch in thickness. These crystals form on this plateau during warm days when the sun's heat leads to a gentle upward streaming of the cold air with a small amount of moisture in it from beneath. Under these influences combined with the thawing of the surface snow, these large and beautiful ice crystals form rapidly in a single day. We observe that after every still sunny day a crop of these crystals develops on the surface of the névé and remains there until the next wind blows them off. They form a layer about half an inch in thickness over the top of the névé. In the bright sunlight the névé, covered with these sheets of bright reflecting ice crystals, glittered like a sea of diamonds. The heavy runners of our sledge rustled gently as they crushed the crystals by the thousand. It seemed a sacrilege. The sastrugi were large and high, and our sledge bumped very heavily over them with a prodigious rattling of our aluminium cooking gear. It was clear that the blizzard winds blow over this part of the plateau at times with great violence. Apparently all the winds in this quarter, strong enough to form sastrugi, blow from south or west of south or from the south-east. Our run today was twelve miles one hundred and fifty yards.

January 15 – We were up today at six a.m. and found a cold southerly breeze blowing, the temperature being minus 19° Fahrenheit. at 6.30 a.m. Mawson got a good latitude determination today: 72°42'.

At about twenty minutes before true noon Mawson took magnetic observations with the dip circle, and found the angle now only fifteen minutes off the vertical, the dip being 89° 45'. We were very much rejoiced to find we were now so close to the magnetic pole. The observations made by Bernacchi during the two years of the *Discovery* Expedition's sojourn at their winter quarters on Ross Island showed the amplitude of daily swing of the magnet was sometimes considerable. The compass, at a distance from the pole, pointing in a slightly varying direction at different times of the day, indicates that the polar centre executes a daily round of wanderings about its mean position. Mawson considered that we were now practically at the magnetic pole, and that if we were to wait for twenty-four hours taking constant observations at this spot the pole would probably during that time come vertically beneath us. We decided, however, to go on to the spot where he concluded the approximate mean position of the magnetic pole would lie. That evening the dip was 89°48'. The run for the day was fourteen miles.

From the rapid rate at which the dip had been increasing recently, as well as from a comparison of Bernacchi's magnetic observations, Mawson estimated we were now about thirteen miles distant from the probable mean position of the south magnetic pole. He stated that in order to accurately locate the mean position possibly a month of continuous observation would be needed, but that the position he indicated was now as close as we could locate it. We decided accordingly, after discussing the matter fully that night, to make a forced march of thirteen miles to the approximate mean position of the pole on the following day, put up the flag there, and return eleven miles back on our tracks the same day. Our method of procedure on this journey of twenty-four miles is described in the journal of the following day.

Saturday, January 16 – We were up at about six a.m. and after breakfast we pulled on our sledge for two miles. We then depoted all our heavy gear and equipment with the exception of the tent, sleeping bag, primus stove and cooker and a small quantity of food, all of which we placed on the sledge, together with the legs of the dip circle and those of the theodolite to serve as marks. We pulled on for two miles and fixed up the legs of the dip circle to guide us back on our track, the compass moving in a horizontal plane being now useless for keeping us on our course.

At two miles further we fixed up the legs of the theodolite, and two miles further put up our tent and had a light lunch. We then walked five miles in the direction of the magnetic pole so as to place us in the mean position calculated for it by Mawson: 72°25' South latitude, 155°16' East longitude. Mawson placed his camera so as to focus

Scottish doctor Alistair Mackay and Australian geologists Edgeworth David and Douglas Mawson pose for the camera – operated by a handheld string – taking "possession of this area now containing the magnetic pole for the British Empire", January 16, 1909.

the whole group, and arranged a trigger which could be released by means of a string held in our hands so as to make the exposure by means of the focal plane shutter. Meanwhile Mackay and I fixed up the flagpole. We then bared our heads and hoisted the Union Jack at 3.30 p.m. with the words uttered by myself, in conformity with Lieutenant Shackleton's instructions, "I hereby take possession of this area now containing the magnetic pole for the British Empire." At the same time I fired the trigger of the camera by pulling the string. Thus the group were photographed in the manner shown on the plate. The blurred line connected with my right hand represents the part of the string in focus blown from side to side by the wind. Then we gave three cheers for his Majesty the King.

There was a pretty sky at the time to the north of us with low cumulus clouds, and we speculated at the time as to whether it was possible that an arm of the sea, such as would produce the moisture to form the cumulus, might not be very far distant. In view of our subsequent discovery of a deep indent in the coastline in a southerly direction beyond Cape North, it is possible the sea at this point is at no very considerable distance.

The temperature at the time we hoisted the flag was exactly 0° Fahrenheit. It was an intense satisfaction and relief to all of us to feel that at last, after so many days of toil, hardship and danger, we had been able to carry out our leader's instructions, and to fulfil the wish of Sir James Clarke Ross that the south magnetic pole should be actually reached, as he had already in 1831 reached the north magnetic pole. At the same time we were too utterly weary to be capable of any great amount of exultation. I am sure the feeling that was uppermost in all of us was one of devout and heartfelt thankfulness to the kind providence which had so far guided our footsteps in safety to that goal.

With a fervent "Thank God" we all did a right-about turn and as quick a march as tired limbs would allow back in the direction of our little green tent in the wilderness of snow.

It was a weary tramp back over the hard and high sastrugi and we were very thankful when at last we saw a small dark cone, which we knew was our tent, rising from above the distant snow ridges. On reaching the tent we each had a little cocoa, a biscuit and a small lump of chocolate. We then sledged slowly and wearily back, picking up first the legs of the theodolite, then those of the dip circle, and finally reached our depot a little before ten p.m.

In honour of the event we treated ourselves that night to a hoosh, which though modest was larger in volume than usual, and was immensely enjoyed. Mawson repacked the sledge after hoosh time, and we turned into the sleeping bag faint and weary, but happy with the great load of apprehension of possible failure, that had been hanging over us for so many weeks, at last removed from our minds. We all slept soundly after twenty-four miles of travel.

Earth's magnetic field is not static, and the south magnetic pole is now located in the ocean north of the polar circle.

From "An Account of the First Journey to the South Magnetic Pole" by Edgeworth David, in *The Heart of the Antarctic*, edited by Ernest Shackleton, volume two: Heinemann, 1909.

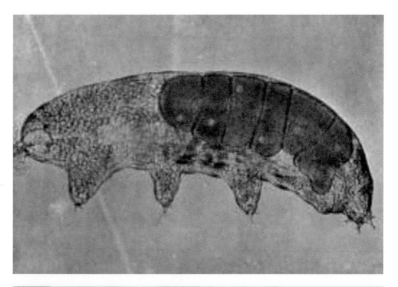

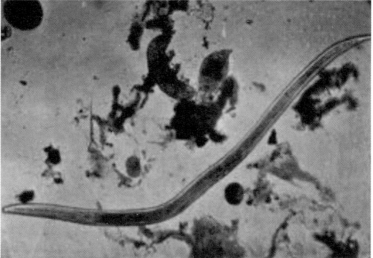

Photographs of a tardigrade, or water bear, (top) and a nematode worm, microscopic creatures James Murray found living under extreme conditions at Cape Royds.

Nematodes, rotifers,
water bears and mites

James Murray was a biologist on Ernest Shackleton's 1907– 09 *Nimrod* Expedition and in charge of the base at Cape Royds, thirty-seven kilometres north of Hut Point on Ross Island and near an Adélie penguin colony. His first impression of the area, with its barren ridges of black lava, moraine and snowdrifts, was that it was "difficult to imagine a more unpromising field for biological study" but he soon changed his mind.

Before his trip to Antarctica, Murray had researched the microscopic life of Scottish lakes. His specialty was tardigrades – eight-legged microscopic animals also known as water bears because of their pudgy appearance when viewed under a microscope. It wasn't until the late twentieth century that the extent of extremophile life in Antarctica was revealed but Murray was a pioneer, finding many creatures capable of surviving Antarctica's extreme cold and aridity. As well as tardigrades, he found other microscopic creatures such as rotifers, rhizopods and nematode worms surviving in the ice-cold lakes and mosses.

He starts this passage from his scientific report on the expedition by comparing the microscope life at Cape Royds to creatures he's familiar with from his work in the northern hemisphere.

The finding of an abundant microscopic fauna and flora at Cape Royds came somewhat as a surprise. It is true that the most northerly lands hitherto carefully examined ... have a rich microscopic fauna, but in these lands the higher summer temperature allows of a flora of the higher plants, and a luxuriant growth of mosses, among which so many microscopic animals have their haunts. ...

111

Cape Royds, though at a much lower latitude (77°30'S) and close by the open sea, has a much lower summer temperature. The mean temperature of a summer day rarely rises above freezing point, and there is no vegetation higher than mosses. As contrasted with the northern lands the moss fauna is a very poor one. We found only four species, and from the whole of Victoria Land there are but eight species known. At Cape Royds they are very scarce, and are stunted and sickly in growth. The micro-fauna which they support is very meagre, a few water bears and rotifers, one rhizopod, and little else. In some tufts of moss the individual animals were numerous; in others no life could be detected.

The kinds of animals which are usually to be found among mosses have at Cape Royds a shelter of another sort, which, judging from their numbers, appears to suit them better. This is furnished by the foliaceous vegetation which grows so abundantly in the lakes and ponds. On the surface and between the layers of this plant they abound both summer and winter. In summer, when the ponds are melted, they enjoy for some weeks a warm climate, the temperature rising as high as 60°F in some ponds. There they are sheltered from the air, which would freeze them every day if they lived among the mosses. In winter again they are frozen in the ice for many months, in some of the deeper lakes for many years.

While the mosses appear to be dwarfed by the cold, the microscopic animals are not at all troubled by the rigours of the climate. When the cold comes they curl up and go to sleep, it may be for years, and when the thaw comes they go merrily on as though nothing had happened. Indeed, since the cold does not harm them, the ice preserves them, secure against all other dangers, except only the advent of explorers. Their numbers prove how completely they are adapted to the local conditions. I have never anywhere seen bdelloid rotifers so plentiful as are the two dominant species at Cape Royds (*Philodina gregaria* and *Adineta grandis*).

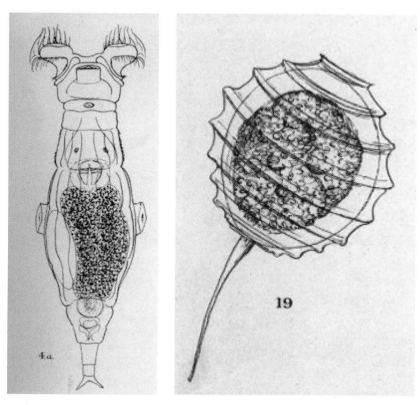

Drawings of a rotifer, *Philodina gregaria* (left), and a flagellatum found in lakes and mosses at Cape Royds. "While the mosses appear to be dwarfed by the cold, [these] microscopic animals are not at all troubled by the rigours of the climate," Murray observed.

Among the higher Invertebrata the rotifers are easily first in numbers, both of individuals and species. The water bears are of only a few kinds, but one of them (*Macrobiotus arcticus*) is extremely abundant. There are nematode worms of two or more kinds, mites of several kinds, and two crustacea belonging to the Entomostraca. The Ciliate Infusoria are very numerous, there are a good many Flagellata, but only two rhizopods were observed.

The vegetation consists solely of algae, blue-green and green, in filaments, colonies, and single cells. The diatoms are few and very small, and the desmids very rare, only two filamentous kinds being seen.

Mites. In Coast Lake, Blue Lake, Clear Lake, and Deep Lake (near Cape Barne) skins of small mites were got. During our stay in the Antarctic none were seen alive, but after our return to England a living mite was obtained from vegetation brought from Deep Lake. It had probably hatched from an egg. In Coast Lake they were abundant and of several species. No drawings or photographs were made but a specimen of one kind was mounted, and it is hoped that enough examples will yet be got to enable us to report upon them.

Insect. The only photograph we obtained of an Antarctic insect was of a parasite on a MacCormick's skua. Two examples were got by Joyce on one skua, and it appears to be rare as [a] very careful search failed to find any others. On an emperor penguin also a small louse was seen, but the specimen was lost.

Other arthropods. In the lakes we occasionally found fragments of arthropods, but whether of crustacean, insect or acarid we did not find out. A probable explanation of such occurrences is that they were parts of small marine crustacea, brought by penguins as food for their chicks and blown into the lakes.

Worms. In addition to the Rotifera we found worms belonging to several other groups – Gastrotricha, Turbellaria, and Nematoda.

Gastrotricha: This small group, supposed to stand near the Rotifera, was represented by a single example of Choetonotus found among weed from the Deep Lake at Cape Barne.

Turbellaria: Microscopic Turbellaria were found in Coast Lake and Blue Lake. In the former they were at one time very abundant.

Nematoda: The nematodes of the lakes were microscopic and free-living. Two kinds were common, one of which is figured on Plate IV, Fig. 13 [page 110, bottom]. Another kind had the skin minutely annulate.

Infusoria. In dealing with the lowest and simplest forms of life, which are easily carried about in the form of dust, there must always be some doubt as to whether many of the species are native or introduced by the expedition making the observations.

When we arrived at Cape Royds the season was well advanced towards autumn, and nearly all the lakes were already frozen. When we cut out blocks of ice containing portions of the vegetation which so abounds in these lakes we found many kinds of Infusoria, some of them of large size, dead and embedded in the ice. These were undoubtedly native. Afterwards, when the lakes melted and living Infusoria appeared in them, we were able to recognise many of them as of the same kinds which we had previously found frozen into the ice. Some of them were encysted and probably alive when found in the ice, but we never observed any of them leaving the cysts while under observation.

One of the puzzling organisms which we first observed consisted of clusters of whitish elliptical bodies, in which no definite organs could be seen. They were supposed to be some kind of eggs. Long afterwards they were accidentally discovered to be vorticellids. On treating with formalin an "infusion" in which a kind of vorticellid abounded, it was found that they contracted into the puzzling egg-like bodies.

There is little doubt that most, if not all, of the Infusoria and other organisms hereafter figured were true natives of the lakes of Cape Royds. The number of kinds seen was much greater than the number noted and figured. Very many were seen at times when important observations were in progress, which allowed of no time being given to side issues. The Flagellata, on account of the greater difficulty attending their study, were generally passed over without note.

Rhizopoda. The paucity of rhizopods at Cape Royds was surprising, after it became known that so many other kinds of microscopic life abounded in the lakes.

In the lakes only two testaceous species were observed: the well-known *Difflugia vas*, and a very small kind which appears to be a Quadrula. Among the moss there was another species, not identified.

On one occasion, when the ice of a lake was melted, we found numbers of an amorphous granular organism, each with a round nucleus, which were probably amoebae, but being only seen dead nothing could be made of them.

In the lakes they also found heliozoa, algae, bacteria and what Murray referred to as "organisms of doubtful position".

From "Part II: Microscopic Life at Cape Royds" by James Murray, in *British Antarctic Expedition 1907–9, Reports on the Scientific Investigations*, volume one: Heinemann, London, 1910.

The measure

And so it proved, cried, grunted,
One man says glass, one ice, one ivory

In this year since our ship got stuck
we measured each second, its sinking
we recorded its fall

And after it sank we walked to a rock valley
where ice spilled like flooded milk between knees
and we were thirsty

And so it proved, cried, grunted,
One man says glass, one ice, one ivory

And we measured the red
of our lips
as we knelt

and we drank every neuron
from the brain of the lake

Alice Miller

These three emperor penguin eggs, collected by Edward Wilson, Henry Bowers and Apsley Cherry-Garrard at Cape Crozier, were given to the Natural History Museum in London by Cherry-Garrard, the only one of the three to make it back to England alive.

The worst journey in the world

Apsley Cherry-Garrard was just twenty-four when he left
England with Robert Falcon Scott on the second of Scott's
Antarctic expeditions. The *Terra Nova* departed from Cardiff in
June 1910, took on supplies – including food, thirty-four dogs,
nineteen Siberian ponies and three motorised sledges – at
Port Chalmers in New Zealand in November, and landed at
Cape Evans in January 1911.

The twin goals of the expedition were to make scientific
discoveries and reach the South Pole. As assistant biologist,
Cherry-Garrard took part in a mission to recover an egg of the
emperor penguin for science. The colony they headed to was
at Cape Crozier, the opposite end of Ross Island from the base
at Hut Point, and had been first sighted during the *Discovery*
Expedition of 1901–04. Cherry-Garrard's two companions on
the winter journey were biologist Edward "Bill" Wilson and
Henry "Birdie" Bowers. Both would die less than a year later,
on the return journey from the South Pole with Scott.

In this extract from his book *The Worst Journey in the World*,
"pressure" is Cherry-Garrard's shorthand for the pressure
ridges formed from collisions between ice floes. These ridges
made dragging a sledge very difficult, especially in the dark.
The "Barrier" is the Great Ice Barrier, the name Ross gave
to the edge of what is known today as the Ross Ice Shelf.

What is this venture? Why is the embryo of the emperor penguin so
important to science? And why should three sane and common sense
explorers be sledging away on a winter's night to a cape which has
only been visited before in daylight, and then with very great difficulty?

The emperor is a bird which cannot fly, lives on fish and never
steps on land, even to breed. It lays its eggs on the bare ice during the

winter and carries out the whole process of incubation on the sea ice, resting the egg upon its feet, pressed closely to the lower abdomen. But it is because the emperor is probably the most primitive bird in existence that the working out of his embryology is so important. The embryo shows remains of the development of an animal in former ages and former states; it recapitulates its former lives. The embryo of an emperor may prove the missing link between birds and the reptiles from which birds have sprung.

Only one rookery of emperor penguins had been found at this date, and this was on the sea ice inside a little bay of the Barrier edge of Cape Crozier, which was guarded by miles of some of the biggest pressure in the Antarctic. Chicks had been found in September, and Wilson reckoned that the eggs must be laid in the beginning of July. And so we started just after midwinter on the weirdest bird's-nesting expedition that has ever been or ever will be. ...

Pulling out from Hut Point that evening we brought along our heavy loads on the two nine-foot sledges with comparative ease; it was the first, and though we did not know it then, the only bit of good pulling we were to have. Good pulling to the sledge traveller means easy pulling. Away we went round Cape Armitage and eastwards. We knew that the Barrier edge was in front of us and also that the break-up of the sea ice had left the face of it as a low perpendicular cliff. We had therefore to find a place where the snow had formed a drift. This we came right up against and met quite suddenly a very keen wind flowing, as it always does, from the cold Barrier down to the comparatively warm sea ice. The temperature was –47°F, and I was a fool to take my hands out of my mitts to haul on the ropes to bring the sledges up. I started away from the Barrier edge with all ten fingers frostbitten. They did not really come back until we were in the tent for our night meal, and within a few hours there were two or three large blisters, up to an inch long, on all of them. For many days those blisters hurt frightfully.

We were camped that night about half a mile in from the Barrier edge. The temperature was −56°. We had a baddish time, being very glad to get out of our shivering bags next morning (29 June). We began to suspect, as we knew only too well later, that the only good time of the twenty-four hours was breakfast, for then with reasonable luck we need not get into our sleeping bags again for another seventeen hours.

The horror of the nineteen days it took us to travel from Cape Evans to Cape Crozier would have to be re-experienced to be appreciated; and anyone would be a fool who went again: it is not possible to describe it. The weeks which followed them were comparative bliss, not because later our conditions were better – they were far worse – but because we were callous. I for one had come to that point of suffering at which I did not really care if only I could die without much pain. They talk of the heroism of the dying – they little know it would be so easy to die, a dose of morphia, a friendly crevasse, and blissful sleep. The trouble is to go on ...

It was the darkness that did it. I don't believe minus seventy temperatures would be bad in daylight, not comparatively bad, when you could see where you were going, where you were stepping, where the sledge straps were, the cooker, the primus, the food; could see your footsteps lately trodden deep into the soft snow [so] you might find your way back to the rest of your load; could see the lashings of the food bags; could read a compass without striking three or four different boxes to find one dry match; could read your watch to see if the blissful moment of getting out of your bag was come without groping in the snow all about; when it would not take you five minutes to lash up the door of the tent, and five hours to get started in the morning ...

But in these days we were never less than four hours from the moment when Bill cried "Time to get up" to the time when we got into our harness. It took two men to get one man into his harness and was all they could do, for the canvas was frozen and our clothes were frozen until sometimes not even two men could bend them into the required shape.

The trouble is sweat and breath. I never knew before how much of the body's waste comes out through the pores of the skin. On the most bitter days, when we had to camp before we had done a four-hour march in order to nurse back our frozen feet, it seemed that we must be sweating. And all this sweat, instead of passing away through the porous wool of our clothing and gradually drying off us, froze and accumulated. It passed just away from our flesh and then became ice: we shook plenty of snow and ice down from inside our trousers every time we changed our foot gear, and we could have shaken it from our vests and from between our vests and shirts, but of course we could not strip to this extent. But when we got into our sleeping bags, if we were fortunate, we became warm enough during the night to thaw this ice: part remained in our clothes, part passed into the skins of our sleeping bags, and soon both were sheets of armour plate.

As for our breath – in the daytime it did nothing worse than cover the lower parts of our faces with ice and solder our balaclavas tightly to our heads. It was no good trying to get your balaclava off until you had had the primus going quite a long time, and then you could throw your breath about if you wished. The trouble really began in your sleeping bag, for it was far too cold to keep a hole open through which to breathe. So all night long our breath froze into the skins, and our respiration became quicker and quicker as the air in our bags got fouler and fouler: it was never possible to make a match strike or burn inside our bags!

Of course we were not iced up all at once: it took several days of this kind of thing before we really got into big difficulties on this score. It was not until I got out of the tent one morning fully ready to pack the sledge that I realised the possibilities ahead. We had had our breakfast, struggled into our foot gear, and squared up inside the tent, which was comparatively warm. Once outside, I raised my head to look round and found I could not move it back. My clothing had frozen hard as I stood – perhaps fifteen seconds. For four hours I had to pull with my head stuck up, and from that time we all took care to bend down into a pulling position before being frozen in.

By now we had realised that we must reverse the usual sledging routine and do everything slowly, wearing when possible the fur mitts which fitted over our woollen mitts, and always stopping whatever we were doing directly we felt that any part of us was getting frozen, until the circulation was restored. Henceforward it was common for one or other of us to leave the other two to continue the camp work while he stamped about in the snow, beat his arms, or nursed some exposed part. But we could not restore the circulation of our feet like this – the only way then was to camp and get some hot water into ourselves before we took our foot gear off. The difficulty was to know whether our feet were frozen or not, for the only thing we knew for certain was that we had lost all feeling in them. Wilson's knowledge as a doctor came in here: many a time he had to decide from our descriptions of our feet whether to camp or to go on for another hour. A wrong decision meant disaster, for if one of us had been crippled the whole party would have been placed in great difficulties. Probably we should all have died.

> **Cherry-Garrard and his companions continued travelling east through the cold and blizzards for more than three weeks. The three men dragged two sledges carrying a total load of more than 300 kilograms.**

And then we heard the emperors calling.

Their cries came to us from the sea ice we could not see, but which must have been a chaotic quarter of a mile away. They came echoing back from the cliffs as we stood helpless and tantalised. We listened and realised that there was nothing for it but to return, for the little light which now came in the middle of the day was going fast, and to be caught in absolute darkness there was a horrible idea. We started back on our tracks and almost immediately I lost my footing and rolled down a slope into a crevasse. Birdie and Bill kept their balance and I clambered back to them. The tracks were very faint and we soon began to lose them. Birdie was the best man at following tracks I have ever known, and he found them time after time. But at last even he lost them altogether and we settled we must just go ahead. As a matter of fact, we picked them up again, and by then were out of the worst: but we were glad to see the tent.

The next morning (Thursday, 20 June) we started work on the igloo at three a.m. and managed to get the canvas roof on, in spite of a wind which harried us all that day. Little did we think what that roof had in store for us as we packed it in with snow blocks, stretching it over our second sledge, which we put athwartships across the middle of the longer walls. The windward (south) end came right down to the ground and we tied it securely to rocks before packing it in. On the other three sides we had a good two feet or more of slack all round, and in every case we tied it to rocks by lanyards at intervals of two feet. The door was the difficulty, and for the present we left the cloth arching over the stones, forming a kind of portico. The whole was well packed in and over with slabs of hard snow, but there was no soft snow with which to fill up the gaps between the blocks. However, we felt already that nothing could drag that roof out of its packing, and subsequent events proved that we were right.

It was a bleak job for three o'clock in the morning before breakfast, and we were glad to get back to the tent and a meal for we meant to have another go at the emperors that day. With the first glimpse of light we were off for the rookery again.

But we now knew one or two things about that pressure which we had not known twenty-four hours ago; for instance, that there was a lot of alteration since the *Discovery* days and that probably the pressure was bigger. As a matter of fact it has been since proved by photographs that the ridges now ran out three-quarters of a mile farther into the sea than they did ten years before. We knew also that if we entered the pressure at the only place where the ice cliffs came down to the level of the Barrier, as we did yesterday, we could neither penetrate to the rookery nor get in under the cliffs where formerly a possible way had been found. There was only one other thing to do – to go over the cliff. And this was what we proposed to try and do.

Now these ice cliffs are some two hundred feet high and I felt uncomfortable, especially in the dark. But as we came back the day before we had noticed at one place a break in the cliffs from which there hung a snowdrift. It might be possible to get down that drift.

And so, all harnessed to the sledge, with Bill on a long lead out in front and Birdie and myself checking the sledge behind, we started down the slope which ended in the cliff, which of course we could not see. We crossed a number of small crevasses, and soon we knew we must be nearly there. Twice we crept up to the edge of the cliff with no success, and then we found the slope: more, we got down it without great difficulty and it brought us out just where we wanted to be, between the land cliffs and the pressure.

Then began the most exciting climb among the pressure that you can imagine. At first very much as it was the day before – pulling ourselves and one another up ridges, slithering down slopes, tumbling into and out of crevasses and holes of all sorts, we made our way along under the cliffs, which rose higher and higher above us as we neared the black lava precipices which form Cape Crozier itself. We straddled along the top of a snow ridge with a razor-backed edge, balancing the sledge between us as we wriggled: on our right was a drop of great depth with crevasses at the bottom; on our left was a smaller drop also crevassed. We crawled along, and I can tell you it was exciting work in the more than half darkness. At the end was a series of slopes full of crevasses, and finally we got right in under the rock on to moraine, and here we had to leave the sledge.

We roped up and started to worry along under the cliffs, which had now changed from ice to rock and rose 800 feet above us. The tumult of pressure which climbed against them showed no order here. Four hundred miles of moving ice behind it had just tossed and twisted those giant ridges until Job himself would have lacked words to reproach their Maker. We scrambled over and under, hanging on with our axes and cutting steps where we could not find a foothold with our crampons. And always we got towards the emperor penguins, and it really began to look as if we were going to do it this time, when we came up against a wall of ice which a single glance told us we could never cross. One of the largest pressure ridges had been thrown, end on, against the cliff. We seemed to be stopped when Bill found a black hole, something like a fox's earth, disappearing into the bowels of the ice. We looked at it: "Well, here goes!" he said, and put his head in and disappeared. Bowers likewise. It was a longish way, but quite possible to wriggle along, and presently I found myself looking out of the other side with a deep gully below me, the rock face on one hand and the ice on the other. "Put your back against the ice and your feet against the rock and lever yourself along," said Bill, who was already standing on firm ice at the far end in a snow pit. We cut some fifteen steps to get out of that hole. Excited by now and thoroughly enjoying ourselves, we found the way ahead easier, until the penguins' call reached us again and we stood, three crystallised ragamuffins, above the emperors' home. They were there all right, and we were going to reach them, but where were all the thousands of which we had heard?

We stood on an ice foot which was really a dwarf cliff some twelve feet high, and the sea ice, with a good many ice blocks strewn upon it, lay below. The cliff dropped straight, with a bit of an overhang and no snowdrift. This may have been because the sea had only frozen recently; whatever the reason ... it meant that we should have a lot of difficulty in getting up again without help. It was decided that someone must stop on the top with the alpine rope, and clearly that one should be I, for with short sight and fogged spectacles which I could not wear I was much the least useful of the party for the job immediately ahead. Had we had the sledge we could have used it as a ladder, but of course we had left this at the beginning of the moraine miles back.

We saw the emperors standing all together huddled under the Barrier cliff some hundreds of yards away. The little light was going fast: we were much more excited about the approach of complete darkness and the look of wind in the south than we were about our triumph. After indescribable effort and hardship we were witnessing a marvel of the natural world, and we were the first and only men who had ever done so; we had within our grasp material which might prove of the utmost importance to science; we were turning theories into facts with every observation we made – and we had but a moment to give.

The disturbed emperors made a tremendous row, trumpeting with their curious metallic voices. There was no doubt they had eggs, for they tried to shuffle along the ground without losing them off their feet. But when they were hustled a good many eggs were dropped and left lying on the ice, and some of these were quickly picked up by eggless emperors who had probably been waiting a long time for the opportunity. In these poor birds the maternal side seems to have necessarily swamped the other functions of life. Such is the struggle for existence that they can only live by a glut of maternity, and it would be interesting to know whether such a life leads to happiness or satisfaction.

When the men reached the emperor penguins, Wilson made this sketch of the scene. The rookery was on sea ice at Cape Crozier, the destination of their winter journey.

The men of the *Discovery* found this rookery where we now stood. They made journeys in the early spring but never arrived early enough to get eggs and only found parents and chicks. They concluded that the emperor was an impossible kind of bird who, for some reason or other, nests in the middle of the Antarctic winter with the temperature anywhere below seventy degrees of frost and the blizzards blowing, always blowing, against his devoted back. And they found him holding his precious chick balanced upon his big feet, and pressing it maternally, or paternally (for both sexes squabble for the privilege) against a bald patch in his breast. And when at last he simply must go and eat something in the open leads near by, he just puts the child down on the ice, and twenty chickless emperors rush to pick it up. And they fight over it, and so tear it that sometimes it will die. And, if it can, it will crawl into any ice crack to escape from so much kindness, and there it will freeze. Likewise many broken and addled eggs were found, and it is clear that the mortality is very great. But some survive and summer comes; and when a big blizzard is going to blow (they know all about the weather), the parents take the children out for miles across the sea ice until they reach the threshold of the open sea. And there they sit until the wind comes, and the swell rises and breaks that ice floe off, and away they go in the blinding drift to join the main pack ice, with a private yacht all to themselves.

You must agree that a bird like this is an interesting beast, and when, seven months ago, we rowed a boat under those great black cliffs and found a disconsolate emperor chick still in the down, we knew definitely why the emperor has to nest in midwinter. For if a June egg was still without feathers in the beginning of January, the same egg laid in the summer would leave its produce without practical covering for the following winter. Thus the emperor penguin is compelled to undertake all kinds of hardships because his children insist on developing so slowly, very much as we are tied in our human relationships for the same reason. It is of interest that such a primitive bird should have so long a childhood.

But interesting as the life history of these birds must be, we had not travelled for three weeks to see them sitting on their eggs. We wanted the embryos, and we wanted them as young as possible, and fresh and unfrozen, that specialists at home might cut them into microscopic sections and learn from the previous history of birds throughout the evolutionary ages. And so Bill and Birdie rapidly collected five eggs, which we hoped to carry safely in our fur mitts to our igloo upon Mount Terror, where we could pickle them in the alcohol we had brought for the purpose. We also wanted oil for our blubber stove, and they killed and skinned three birds – an emperor weighs up to six and a half stones.

The Ross Sea was frozen over and there were no seals in sight. There were only 100 emperors as compared with 2,000 in 1902 and 1903. Bill reckoned that every fourth or fifth bird had an egg but this was only a rough estimate, for we did not want to disturb them unnecessarily. It is a mystery why there should have been so few birds, but it certainly looked as though the ice had not formed very long. Were these the first arrivals? Had a previous rookery been blown out to sea and was this the beginning of a second attempt? Is this bay of sea ice becoming unsafe?

Those who previously discovered the emperors with their chicks saw the penguins nursing dead and frozen chicks if they were unable to obtain a live one. They also found decomposed eggs which they must have incubated after they had been frozen. Now we found that these birds were so anxious to sit on something that some of those which had no eggs were sitting on ice! Several times Bill and Birdie picked up eggs to find them lumps of ice, rounded and about the right size, dirty and hard. Once a bird dropped an ice nest egg as they watched, and again a bird returned and tucked another into itself – immediately forsaking it for a real one, however, when one was offered.

Meanwhile a whole procession of emperors came round under the cliff on which I stood. The light was already very bad and it was well that my companions were quick in returning: we had to do everything in a great hurry. I hauled up the eggs in their mitts (which we fastened together round our necks with lampwick lanyards) and then the skins, but failed to help Bill at all. "Pull," he cried, from the bottom. "I am pulling," I said. "But the line's quite slack down here," he shouted. And when he had reached the top by climbing up on Bowers' shoulders, and we were both pulling all we knew, Birdie's end of the rope was still slack in his hands. Directly we put on a strain the rope cut into

Wilson, Bowers and Cherry-Garrard, survivors of the tortuous winter expedition to collect emperor penguin eggs for science, eating a meal after making it back from Cape Crozier, August 1, 1911.

the ice edge and jammed – a very common difficulty when working among crevasses. We tried to run the rope over an ice axe without success, and things began to look serious when Birdie, who had been running about prospecting and had meanwhile put one leg through a crack into the sea, found a place where the cliff did not overhang. He cut steps for himself, we hauled, and at last we were all together on the top – his foot being by now surrounded by a solid mass of ice.

We legged it back as hard as we could go: five eggs in our fur mitts, Birdie with two skins tied to him and trailing behind, and myself with one. We were roped up, and climbing the ridges and getting through the holes was very difficult. In one place where there was a steep rubble and snow slope down, I left the ice axe halfway up; in another it was too dark to see our former ice-axe footsteps, and I could see nothing, and so just let myself go and trusted to luck. With infinite patience Bill said, 'Cherry, you must learn how to use an ice axe.' For the rest of the trip my wind-clothes were in rags.

We found the sledge and none too soon, and now had three eggs left, more or less whole. Both mine had burst in my mitts: the first I emptied out, the second I left in my mitt to put into the cooker; it never got there, but on the return journey I had my mitts far more easily thawed out than Birdie's (Bill had none) and I believe the grease in the egg did them good. When we got into the hollows under the ridge where we had to cross, it was too dark to do anything but feel our way. We did so over many crevasses, found the ridge and crept over it. Higher up we could see more, but to follow our tracks soon became impossible, and we plugged straight ahead and luckily found the slope down which we had come.

All day it had been blowing a nasty cold wind with a temperature between –20° and –30°, which we felt a good deal. Now it began to get worse. The weather was getting thick and things did not look very nice when we started up to find our tent. Soon it was blowing force four, and soon we missed our way entirely. We got right up above the patch of rocks which marked our igloo and only found it after a good deal of search.

> In 1913 Cherry-Garrard, the sole surviving member of the team who braved this 200-kilometre journey, in near darkness and temperatures ranging from −40°C to −60°C, delivered three intact emperor penguin eggs to the chief custodian of the Natural History Museum in London. He later described the rudeness with which he was treated: the indifferent response the eggs elicited made him feel ready to "commit murder".

Although years later he reprimanded a museum custodian that "the specimens brought by the expedition from Antarctica did not include the moths we found preying on some of them", the eggs survived and remain today at the museum.

From *The Worst Journey in the World* by Apsley Cherry-Garrard: Carroll & Graf, London, 1922.

Adélie penguins at the foot of the ice at Cape Adare, photographed by George Murray Levick in 1911.

The hooligan cocks of Cape Adare

The Adélie penguin, *Pygoscelis adeliae*, is often described as the most human-like of all the penguin species. Perhaps that is why this account of its sexual habits was for many years considered too offensive for publication: necrophilia, sexual coercion and homosexuality did not fit with either Darwinian scientific ideas or notions of how a human-like creature would behave.

George Murray Levick, staff surgeon on the British *Terra Nova* Expedition, spent nearly a year at Cape Adare, from February 1911 to January 1912, much of the time photographing penguins and observing their behaviour. After the expedition's return to London, Levick's account of the penguins was published in two separate volumes – in 1914 as *Antarctic Penguins – A Study of Their Social Habits* and in 1915 as *Natural History of the Adélie Penguin*. In the second book Levick talked about "hooligan cocks" and asserted "the crimes which they commit are such as to find no place in this book". A short piece about their wicked behaviour was printed as *The Sexual Habits of the Adélie Penguin* and privately distributed. It was rediscovered in the Natural History Museum at Tring and published in *Polar Record* in 2012.

On their arrival at the rookery, pairing takes place among the penguins in the manner I have described elsewhere. Mated couples copulate very frequently, sometimes more than once a day, not only before the eggs have been laid but for long after, and I have seen a cock copulating with a hen as she sat on her two eggs on the nest, and later after the chicks were well grown.

It is evident that in a vast rookery such as that at Cape Adare, when comparatively few unmated birds are left many of these, scattered in that great crowd, may seek one another for many days, and that some, perhaps, never meet at all.

Owing, no doubt, to the fact that the season had arrived when their sexual functions were at the height of their activity, and in part to the sights and sounds which surrounded them constantly, cocks were often seen whose passions seemed to have passed beyond their control. Sometimes we saw these birds, after walking some distance, apparently in the vain search for hens, stand motionless and rigid upon the ground, then, stiffening themselves, assume the attitude and go through the motions characteristic of the sexual act, in some cases actually ejecting their semen on to the ground.

This, however, was the least depraved of the acts we saw. Strewn about all Antarctic rookeries are the dead bodies of many hundreds of penguins, from the adult to the newly hatched chick, which have succumbed for various reasons during previous years. Owing to the low temperature prevailing, these bodies are preserved in good condition for a long time, several years passing in many cases before they lose their fresh appearance, whilst many of those that have died the year before are still preserved in good plumage.

On November 10th, i.e. when the season was already a month advanced, I saw a cock engaged in the sexual act upon the dead body of a white-throated Adélie of the previous year. This took somewhat over a minute, the position taken up by the cock differing in no way from that of normal copulation, and the whole act was gone through, down to the final depression of the cloaca and emission of semen.

On returning to the hut I told one of my companions what I had seen, and to my surprise he at once said that he had on several occasions seen the same thing done to dead bodies along the ice foot. Later on, this sight was by no means uncommon.

As the season advanced, the number of unmated cocks increased to a great extent, partly from the reasons already given, partly owing to the large number of homes now being broken up owing to accidental destruction of the eggs, depredations of skuas, etc., and also in a large measure to the ravages of the sea leopards gathered in the sea in the vicinity of the rookery. These unmated cocks congregate in little "hooligan" bands of half a dozen or more and hang about the outskirts of the knolls, whose inhabitants they annoy by their constant acts of depravity.

I have said that cramp or some sort of paralysis occasionally attacks the penguins after they have been in the sea. One day I was watching a hen painfully dragging herself across the rookery on her belly, using her flippers for propulsion as her legs trailed uselessly behind her. As I was just wondering whether I ought to kill her or not, a cock, seeing her pass, ran out from the outskirts of a neighbouring knoll and went up to her. After a short inspection he deliberately copulated with her, she being, of course, quite unable to resist him. He had hardly left her before another cock ran up, and without any hesitation tried to mount her. He fell off at first and then, desisting, stole two stones from neighbouring nests, dropped them one after the other in front of her, after which he mounted and performed the sexual act.

When he had gone, the poor hen struggled on about twenty yards and then another cock ran up to her, and was just going to do the same thing when a fourth came up and fought him, driving him away, and afterwards did as the others had done.

After this the hen, who now seemed much more lively, struggled on and had gone about ten yards further when no less than three more cocks gave chase, all trying to climb on to her at once, but this ended in a short fight, after which they went their several ways.

The hen lay still, doubtless being much in need of rest, and as the poor thing evidently knew her way and was making in a straight line I left her, deciding that she might recover if she reached her own nest.

George Murray Levick skinning a penguin on board the *Terra Nova* on the long voyage south, December 28, 1910.

Commander Campbell, whom I called up, witnessed the above scene with me.

Later on I went out and found the hen again. She was much better, and able to stand up and hobble about on her legs. Her back, I am sorry to say, bore signs of further indignities, though she was for the time in peace.

When the chicks are young, the parents take great pains to keep them on their own nests, though occasionally they stray and lose their lives as a result. Very often they suffer indignity and death at the hands of the hooligan cocks, the waste of life resulting from this being very considerable. Frequently we saw strayed chicks sexually misused by these hooligans, some of them being crushed to such an extent that they died in consequence.

On one occasion one of the two chicks which were with their mother on a nest strayed away a short distance and was at once caught by a hooligan cock, who misused it before the very eyes of its parent. The latter kept to her nest and the other chick whilst this took place, and when the strayed chick escaped from the cock and ran back to her she would have nothing to do with it, pecking it whenever it attempted to return to the nest. Eventually it abandoned the attempt and tried to get itself adopted by several other parents, none of which would have it, and it was so severely pecked that I was obliged at length to kill it to put it out of its pain.

The hooligan cocks seemed always to be on the watch for these strayed chicks, and as some of the colonies constantly had a little knot of these hanging about their outskirts a chick, once it had lost itself, was almost certain to come to a speedy end in this way, should it not first be carried off by a skua.

Later in the season, as I have said, a large number of homes were broken up owing to the death of one or other of their occupants. On the southern edge of the rookery, close to the ice foot and away from the nests, was a stretch of basaltic shingle. This became the haunt of a number of desolate mates, a large group of whom were always to be seen there, standing or squatting on the stones.

Many of the hens made scoops in the shingle and lay in them, and though cocks repeatedly had sexual intercourse with them no second attempt was made to form a home, and no nests appeared on the site of the scoops.

Here on one occasion I saw what I took to be a cock copulating with a hen. When he had finished, however, and got off, the apparent hen turned out to be a cock, and the act was again performed with their positions reversed, the original "hen" climbing on to the back of the original cock, whereupon the nature of their proceeding was disclosed.

From "Dr George Murray Levick (1876–1956): unpublished notes on the sexual habits of the Adélie penguin" by Douglas G.D. Russell, William J.L. Sladen and David G. Ainley in *Polar Record*: Cambridge University Press, January 2012.

George Murray Levick and the Adélie penguins

Blubber and soot and
blubber and eyes raw with
the smoke from the blubber fuelled
lamps and clothes stiff from cooking with
blubber. Audrey and the girls
are so far away and the little
hooligans are everywhere
committing their depraved
acts all the long Antarctic
summer. Now I'll have the winter
to wait and think. About young males
and their desperate moves
in penguin suits.

Helen Heath

Robert Falcon Scott writing in his journal in the hut at Cape Evans on Ross Island, main base of his shore party, in 1911, during the *Terra Nova* Expedition that was to claim his life the following year.

Impressions

Robert Falcon Scott kept diaries on both of his Antarctic expeditions. Most of the entries were day-to-day details of his party's travel, meals and state of health, but these notes from his entry of February 2, 1911 read more like a poem. Poet Bill Manhire included this extract in his 2004 anthology *The Wide White Page: Writers Imagine Antarctica.* It describes a short journey from the expedition's base on Ross Island to lay supply depots on the Ross Ice Shelf for the following year's attempt on the South Pole. Scott is travelling with a group of twelve men, eight ponies and twenty-six dogs.

The seductive folds of the sleeping bag.

The hiss of the primus and the fragrant steam of the cooker issuing from the tent ventilator.

The small green tent and the great white road.

The whine of a dog and the neigh of our steeds.

The driving cloud of powdered snow.

The crunch of footsteps which break the surface crust.

The wind-blown furrows.

The blue arch beneath the smoky cloud.

The crisp ring of the ponies' hoofs and the swish of the following sledge.

The droning conversation of the march as driver encourages or chides his horse.

The patter of dog pads.

The gentle flutter of our canvas shelter.

Its deep booming sound under the full force of a blizzard.

The drift snow like finest flour penetrating every hole and corner –

A photograph taken by Scott on December 2, 1911 of ponies pulling sleds on the Ross Ice Shelf, led by members of his expedition.

flickering up beneath one's head covering, pricking sharply as a sand blast.

The sun with blurred image peeping shyly through the wreathing drift giving pale shadowless light.

The eternal silence of the great white desert. Cloudy columns of snowdrift advancing from the south, pale yellow wraiths, heralding the coming storm, blotting out one by one the sharp-cut lines of the land.

The blizzard, Nature's protest – the crevasse, Nature's pitfall – that grim trap for the unwary – no hunter could conceal his snare so perfectly – the light rippled snow bridge gives no hint or sign of the hidden danger, its position unguessable 'til man or beast is floundering, clawing and struggling for foothold on the brink.

The vast silence broken only by the mellow sounds of the marching column.

From *Scott's Last Expedition: The Journals of Captain R.F. Scott* by Robert Falcon Scott: John Murray, London, 1923.

The camp on the Beardmore Glacier set up by Robert Falcon Scott's expedition during their return trek from the South Pole in the summer of 1912. Despite exhaustion and the long and arduous journey, the men continued to collect scientific specimens, including over fifteen kilograms of rocks. Among them was a valuable fossil of a *Glossopteris* fern.

Geologising on the Beardmore

After reaching the South Pole on January 17, 1912 and finding the Norwegian explorer Roald Amundsen had got there one month earlier, Scott and his men had to turn around and face the 1,300-kilometre journey back to Ross Island, where a group of men were waiting for them at Discovery Hut. Scott and his four travelling companions – Lawrence Oates, Henry "Birdie" Bowers, Edward "Bill" Wilson and Edgar Evans – travelled on skis, dragging behind them a sledge loaded with all their food, tents and supplies. As they made the long trek, through unseasonably low temperatures and with dwindling food supplies, their fixed points along the way were the supply depots laid on the forward journey or by other members of the *Terra Nova* Expedition.

In this excerpt from Scott's diary, which was recovered from the tent when he and his men were found the following spring and first published in 1913, they are just coming off the Polar Plateau, down the Beardmore Glacier to the Ross Ice Shelf. Evans is suffering from frostbite, an infected wound on his hand, and a recent fall. All the men are hungry, cold and tired. The temperatures in the narrative are in Fahrenheit and the heights are in feet. The R and the number after it refer to the camps on this return journey.

Tuesday, February 6 – Lunch 7900; Supper 7210. Temp. –15°. R. 20. We've had a horrid day and not covered good mileage. On turning out found sky overcast; a beastly position amidst crevasses. Luckily it cleared just before we started. We went straight for Mt. Darwin, but in half an hour found ourselves amongst huge open chasms, unbridged, but not very deep, I think. We turned to the north between two, but to our chagrin they converged into chaotic disturbance. We had to retrace our steps for a mile or so, then struck to the west and got on to a confused sea of sastrugi, pulling very hard; we put up the sail, Evans' nose suffered, Wilson very cold, everything horrid. Camped for lunch in the sastrugi; the only comfort, things looked clearer to the west and we were obviously going downhill. In the afternoon we struggled on, got out of sastrugi and turned over on glazed surface, crossing many crevasses – very easy work on ski. Towards the end of the march we realised the certainty of maintaining a more or less straight course to the depot, and estimate distance ten to fifteen miles.

Food is low and weather uncertain, so that many hours of the day were anxious; but this evening, though we are not as far advanced as I expected, the outlook is much more promising. Evans is the chief anxiety now; his cuts and wounds suppurate, his nose looks very bad, and altogether he shows considerable signs of being played out. Things may mend for him on the glacier, and his wounds get some respite under warmer conditions. I am indeed glad to think we shall so soon have done with plateau conditions. It took us twenty-seven days to reach the Pole and twenty-one days back – in all forty-eight days – nearly seven weeks in low temperature with almost incessant wind.

End of the summit journey
Wednesday, February 7 – Mount Darwin [or Upper Glacier] Depot, R. 21. Height 7,100. Lunch temp. –9°; Supper temp. [a blank here]. A wretched day with satisfactory ending. First panic, certainty that biscuit-box was short. Great doubt as to how this has come about, as we certainly haven't over-issued allowances. Bowers is dreadfully disturbed about it. The shortage is a full day's allowance. We started our march at 8.30, and travelled down slopes and over terraces covered with hard sastrugi – very tiresome work – and the land didn't seem to come any nearer. At lunch the wind increased, and what with hot tea and good food, we started the afternoon in a better frame of mind,

and it soon became obvious we were nearing our mark. Soon after 6.30 we saw our depot easily and camped next it at 7.30.

Found note from E. Evans to say the second return party passed through safely at 2.30 on January 14 – half a day longer between depots than we have been. The temperature is higher, but there is a cold wind tonight. Well, we have come through our seven weeks' ice camp journey and most of us are fit, but I think another week might have had a very bad effect on P.O. Evans, who is going steadily downhill.

It is satisfactory to recall that these facts give absolute proof of both expeditions having reached the pole and placed the question of priority beyond discussion.

Return from first summit depot
Thursday, February 8 – R. 22. Height 6,260. Start temp. –11°; Lunch temp. –5°; Supper, zero. 9.2 miles. Started from the depot rather late owing to weighing biscuit, &c., and rearranging matters. Had a beastly morning. Wind very strong and cold. Steered in for Mt. Darwin to visit rock. Sent Bowers on, on ski, as Wilson can't wear his at present. He obtained several specimens, all of much the same type, a close-grained granite rock which weathers red. Hence the pink limestone. After he rejoined we skidded downhill pretty fast, leaders on ski, Oates and Wilson on foot alongside sledge – Evans detached. We lunched at two well down towards Mt. Buckley, the wind half a gale and everybody very cold and cheerless.

However, better things were to follow. We decided to steer for the moraine under Mt. Buckley and, pulling with crampons, we crossed some very irregular steep slopes with big crevasses and slid down towards the rocks. The moraine was obviously so interesting that when we had advanced some miles and got out of the wind, I decided to camp and spend the rest of the day geologising. It has been extremely interesting. We found ourselves under perpendicular cliffs of Beacon sandstone, weathering rapidly and carrying veritable coal seams. From the last, Wilson, with his sharp eyes, has picked several plant impressions, the last a piece of coal with beautifully traced leaves in layers, also some excellently preserved impressions of thick stems, showing cellular structure. In one place we saw the cast of small waves in the sand. Tonight Bill has got a specimen of limestone with *archaeocyaths* – the trouble is one cannot imagine where the stone comes from; it is evidently rare, as few specimens occur in the moraine. There is a good deal of pure white quartz. Altogether we have had a most interesting afternoon, and the relief of being out of the wind and in a warmer temperature is inexpressible. I hope and trust we shall all buck up again now that the conditions are more favourable. We have been in shadow all the afternoon, but the sun has just reached us, a little obscured by night haze. A lot could be written on the delight of setting foot on rock after fourteen weeks of snow and ice and nearly seven out of sight of aught else. It is like going ashore after a sea voyage. We deserve a little good bright weather after all our trials, and hope to get a chance to dry our sleeping-bags and generally make our gear more comfortable.

Friday, February 9 – R. 23. Height 5,210 ft. Lunch temp. +10°; Supper temp. +12.5°. About 13 miles. Kept along the edge of moraine to the end of Mt. Buckley. Stopped and geologised. Wilson got great find of vegetable impression in piece of limestone. Too tired to write geological notes. We all felt very slack this morning, partly rise of temperature, partly reaction, no doubt. Ought to have kept close in to glacier north of Mt. Buckley, but in bad light the descent looked steep and we kept out. Evidently we got amongst bad ice pressure and had to come down over an ice-fall. The crevasses were much firmer than expected and we got down with some difficulty, found our night camp of December 20, and lunched an hour after. Did pretty well in the afternoon, marching three and three-quarter hours; the sledgemeter is unshipped, so cannot tell distance traversed. Very warm on march and we are all pretty tired. Tonight it is wonderfully calm and warm, though it has been overcast all the afternoon. It is remarkable to be able to stand outside the tent and sun oneself. Our food satisfies now, but we must march to keep on the full ration, and we want rest, yet we shall pull through all right, D.V. We are by no means worn out.

Saturday, February 10 – R. 24. Lunch temp. +12°; Supper temp. +10°. Got off a good morning march in spite of keeping too far east and getting in rough, cracked ice. Had a splendid night's sleep, showing great change in all faces, so didn't get away 'til ten a.m. Lunched just before three. After lunch the land began to be obscured. We held a course for two and a half hours with difficulty, then the sun disappeared and snow drove in our faces with northerly wind – very warm and impossible to steer, so camped. After supper, still very thick all round, but sun showing and less snow falling. The fallen snow crystals are quite feathery like thistledown. We have two full days' food left, and though our position is uncertain, we are certainly within two outward marches from the middle glacier depot. However, if the weather doesn't clear by tomorrow we must either march blindly on or reduce food. It is very trying. Another night to make up arrears of sleep. The ice crystals that first fell this afternoon were very large. Now the sky is clearer overhead, the temperature has fallen slightly and the crystals are minute.

Sunday, February 11 – R. 25. Lunch temp. −6.5°; Supper temp −3.5°. The worst day we have had during the trip and greatly owing to our own fault. We started on a wretched surface with light S.W. wind, sail set, and pulling on ski – horrible light, which made everything look fantastic. As we went on light got worse, and suddenly we found ourselves in pressure. Then came the fatal decision to steer east. We went on for six hours, hoping to do a good distance, which in fact I suppose we did, but for the last hour or two we pressed on into a regular trap. Getting on to a good surface we did not reduce our lunch meal, and thought all going well, but half an hour after lunch we got into the worst ice mess I have ever been in. For three hours we plunged on on ski, first thinking we were too much to the right, then too much to the left; meanwhile the disturbance got worse and my spirits received a very rude shock.

There were times when it seemed almost impossible to find a way out of the awful turmoil in which we found ourselves. At length, arguing that there must be a way on our left, we plunged in that direction. It got worse, harder, more icy and crevassed. We could not manage our ski and pulled on foot, falling into crevasses every minute – most luckily no bad accident. At length we saw a smoother slope towards the land, pushed for it, but knew it was a woefully long way from us. The turmoil changed in character, irregular crevassed surface giving way to huge chasms, closely packed and most difficult to cross. It was very heavy work, but we had grown desperate. We won through at ten p.m. and I write after twelve hours on the march. I think we are on or about the right track now, but we are still a good number of miles from the depot so we reduced rations tonight. We had three pemmican meals left and decided to make them into four. Tomorrow's lunch must serve for two if we do not make big progress. It was a test of our endurance on the march and our fitness with small supper. We have come through well. A good wind has come down the glacier which is clearing the sky and surface. Pray God the wind holds tomorrow. Short sleep tonight and off first thing, I hope.

Monday, February 12 – R. 26. In a very critical situation. All went well in the forenoon, and we did a good long march over a fair surface. Two hours before lunch we were cheered by the sight of our night camp of the 18th December, the day after we made our depot – this showed we were on the right track. In the afternoon, refreshed by tea, we went forward, confident of covering the remaining distance, but by a fatal chance we kept too far to the left, and then we struck uphill and, tired and despondent, arrived in a horrid maze of crevasses and fissures. Divided councils caused our course to be erratic after this, and finally, at nine p.m. we landed in the worst place of all. After discussion we decided to camp, and here we are, after a very short supper and one meal only remaining in the food bag; the depot doubtful in locality. We *must* get there tomorrow. Meanwhile we are cheerful with an effort. It's a tight place but luckily we've been well fed up to the present. Pray God we have fine weather tomorrow.

Tuesday, February 13 – Camp R. 27, beside Cloudmaker. Temp. –10°. Last night we all slept well in spite of our grave anxieties. For my part these were increased by my visits outside the tent, when I saw the sky gradually closing over and snow beginning to fall. By our ordinary time for getting up it was dense all around us. We could see nothing, and we could only remain in our sleeping bags. At 8.30 I dimly made out the land of the Cloudmaker. At nine we got up, deciding to have tea, and with one biscuit, no pemmican, so as to leave our scanty remaining meal for eventualities. We started marching, and at first had to wind our way through an awful turmoil of broken ice, but in about an hour we hit an old moraine track, brown with dirt. Here the surface was much smoother and improved rapidly. The fog still hung over all and we went on for an hour, checking our bearings. Then the whole plain got smoother and we turned outward a little. Evans raised our hopes with a shout of depot ahead, but it proved to be a shadow on the ice. Then suddenly Wilson saw the actual depot flag. It was an immense relief, and we were soon in possession of our three and a half days' food. The relief to all is inexpressible; needless to say we camped and had a meal. Marching in the afternoon, I kept more to the left, and closed the mountain 'til we fell on the stone moraines. Here Wilson detached himself and made a collection, whilst we pulled the sledge on. We camped late, abreast the lower end of the mountain, and had nearly our usual satisfying supper. Yesterday was the worst experience of the trip and gave a horrid feeling of insecurity. Now we are right, but we must march. In future food must be worked so that we do not run so short if the weather fails us. We mustn't get into a hole like this again. Greatly relieved to find that both the other parties got through safely. Evans seems to have got mixed up with pressures like ourselves. It promises to be a very fine day tomorrow. The valley is gradually clearing. Bowers has had a very bad attack of snow blindness, and Wilson another almost as bad. Evans has no power to assist with camping work.

Wednesday, February 14 – Lunch temp. 0°; Supper temp. +1°. A fine day with wind on and off down the glacier, and we have done a fairly good march. We started a little late and pulled on down the moraine. At first I thought of going right, but soon, luckily, changed my mind and decided to follow the curving lines of the moraines. This course has brought us well out on the glacier. Started on crampons; one hour after, hoisted sail; the combined efforts produced only slow speed, partly due to the sandy snowdrifts similar to those on summit, partly to our torn sledge runners. At lunch these were scraped and sandpapered. After lunch we got on snow, with ice only occasionally showing through. A poor start, but the gradient and wind improving, we did six and a half miles before night camp.

There is no getting away from the fact that we are not pulling strong. Probably none of us: Wilson's leg still troubles him and he doesn't like to trust himself on ski, but the worst case is Evans, who is giving us serious anxiety. This morning he suddenly disclosed a huge blister on his foot. It delayed us on the march, when he had to have his crampon readjusted. Sometimes I fear he is going from bad to worse, but I trust he will pick up again when we come to steady work on ski like this afternoon. He is hungry and so is Wilson. We can't risk opening out our food again, and as cook at present I am serving something under full allowance. We are inclined to get slack and slow with our camping arrangements, and small delays increase. I have talked of the matter tonight and hope for improvement. We cannot do distance without the hours. The next depot some thirty miles away and nearly three days' food in hand.

Thursday, February 15 – R. 29. Lunch temp. +10°; Supper temp. +4°. 13.5 miles. Again we are running short of provision. We don't know our distance from the depot, but imagine about twenty miles. Heavy march – did thirteen and three-quarters (geo.). We are pulling for food and not very strong evidently. In the afternoon it was overcast; land blotted out for a considerable interval. We have reduced food, also sleep; feeling rather done. Trust one and a half days or two at most will see us at depot.

Friday, February 16 – 12.5 m. Lunch temp. +6.1°; Supper temp. +7°. A rather trying position. Evans has nearly broken down in brain, we think. He is absolutely changed from his normal self-reliant self. This morning and this afternoon he stopped the march on some trivial excuse. We are on short rations but not very short, food spins out 'til tomorrow night. We cannot be more than ten or twelve miles from the depot but the weather is all against us. After lunch we were enveloped in a snow sheet, land just looming. Memory should hold the events of a very troublesome march with more troubles ahead. Perhaps all will be well if we can get to our depot tomorrow fairly early, but it is anxious work with the sick man. But it's no use meeting troubles halfway, and our sleep is all too short to write more.

Saturday, February 17 – A very terrible day. Evans looked a little better after a good sleep and declared, as he always did, that he was quite well. He started in his place on the traces, but half an hour later worked his ski shoes adrift and had to leave the sledge. The surface was awful, the soft recently fallen snow clogging the ski and runners at every step, the sledge groaning, the sky overcast, and the land hazy. We stopped after about one hour, and Evans came up again, but very slowly. Half an hour later he dropped out again on the same plea. He asked Bowers to lend him a piece of string. I cautioned him to come on as quickly as he could, and he answered cheerfully as I thought.

We had to push on, and the remainder of us were forced to pull very hard, sweating heavily. Abreast the Monument Rock we stopped, and seeing Evans a long way astern I camped for lunch. There was no alarm at first, and we prepared tea and our own meal, consuming the latter. After lunch, and Evans still not appearing, we looked out, to see him still afar off. By this time we were alarmed and all four started back on ski. I was first to reach the poor man and shocked at his appearance; he was on his knees with clothing disarranged, hands uncovered and frostbitten, and a wild look in his eyes. Asked what was the matter, he replied with a slow speech that he didn't know but thought he must have fainted. We got him on his feet but after two or three steps he sank down again. He showed every sign of complete collapse. Wilson, Bowers, and I went back for the sledge, whilst Oates remained with him. When we returned he was practically unconscious, and when we got him into the tent quite comatose. He died quietly at 12.30 a.m.

On discussing the symptoms we think he began to get weaker just before we reached the pole, and that his downward path was accelerated first by the shock of his frostbitten fingers, and later by falls during rough travelling on the glacier, further by his loss of all confidence in himself. Wilson thinks it certain he must have injured his brain by a fall. It is a terrible thing to lose a companion in this way, but calm reflection shows that there could not have been a better ending to the terrible anxieties of the past week. Discussion of the situation at lunch yesterday shows us what a desperate pass we were in with a sick man on our hands at such a distance from home.

At one a.m. we packed up and came down over the pressure ridges, finding our depot easily.

Things did not go well for the remaining men. On March 16 Oates, who was by now so weak and ill he could no longer contribute to the pulling of the sledge, walked out of the tent and into a blizzard. On March 19, twenty miles further north, the remaining men were confined to the tent by another blizzard. Scott's last diary entry was on March 29.

Next spring a party of his men found his body and those of Wilson and Bowers in a tent just eighteen kilometres from the next depot. They removed diaries and rocks from the tent and sledge, and erected a snow cairn over the bodies.

Scott's day "geologising" at the top of the Beardmore Glacier had added some fifteen kilograms to the load the men had had to haul. Although some people said this extra load may have contributed to their deaths, the rocks were geologically significant. When the remaining members of the expedition returned to England, the rocks were taken to the University of Cambridge. Here Albert Seward, professor of botany, reported they contained fossils of a *Glossopteris* fern, giving new support to the theory of continental drift and the idea that Antarctica had once been part of a giant supercontinent, Gondwana. As a footnote in the 1923 edition of Scott's diary said: "The time spent in collecting, the labour endured in dragging the additional thirty-five pounds to the last camp, were a heavy price to pay. But if the cost was great, the scientific value was great also. The fossils they contain, often so inconspicuous that it is a wonder they were discerned by the collectors, prove to be the most valuable obtained by the expedition, and promise to solve completely the question of the age and past history of this portion of the Antarctic Continent."

From *Scott's Last Expedition: The Journals of Captain R.F. Scott* by Robert Falcon Scott: John Murray, London, 1923.

Plate CCLXXVII: Charles Wright, Griffith Taylor and Raymond Priestley, members of Scott's *Terra Nova* Expedition, were part of a group based at Cape Adare and Terra Nova Bay. Here Griffith and Wright are pictured standing in a grotto within an iceberg, with the ship in the background.

Ice monsters, growlers
and bergy bits

Raymond Priestley's first trip to Antarctica was as geologist
on Shackleton's *Nimrod* Expedition. In 1911 he returned on
the *Terra Nova* Expedition. While Scott's party set up base at
Cape Evans on Ross Island, Priestley joined another party,
led by first officer Victor Campbell and based first at Cape
Adare and then at Terra Nova Bay on the Antarctic continent.
Priestley and expedition physicist Charles Wright published
a book about Antarctic glaciology in 1922, one of the many
scientific accounts of the expedition. This piece about icebergs
was written by Priestley.

Undoubtedly, the polar phenomenon which has most touched the
imagination of sailors of all ages has been the "ice mountain" of the
early navigators – the iceberg of modern nomenclature.

In these days, when steamships are almost universal, the significance
of the iceberg to the crew of the sailing ship is difficult to realise –
though modern disasters such as the loss of the *Titanic* in 1911 serve to
remind us that large masses of floating ice still constitute a menace to
navigation. The Arctic icebergs were known and dreaded for centuries
before those of the south polar region were encountered, and the
marked contrast between the majority of Arctic and Antarctic icebergs
at once struck the early Antarctic navigators, who had nearly all had
experience in Arctic waters, as being very remarkable. The whaling
captains – Captain Cook, Sir James Ross, and in fact most of the early
explorers of Antarctic seas – make special mention of the "tabular"
icebergs of the south as being one of the most characteristic features
of the region. It was not until a near approach to the shores of the

continent was made that the cause of the particular shape of the largest Antarctic icebergs was realised.

The discovery of the Ross Barrier in 1841, and of numerous other similar land ice-formations, at once explained the prevalence of this type of iceberg. They were quite clearly the direct result of that particular stage of the glacial epoch which is associated with the overflow of the ice sheets of the land into the sea, and with a climate and coastal snowfall which involved a sheathing of the continent in locally formed shore ice.

The point which requires emphasising, however, is, that all round the shores of the continent, both where the shores are too steep and exposed to permit the accumulation of sheets of piedmont ice, confluent ice or shelf ice, and too little dissected to allow a sufficient drainage from the continental ice and highland ice sheets to form ice tongues, numerous steep cascading mountain glaciers of the type more familiar to northern eyes pour their quota of ice into the sea. At the snouts of such glaciers and of other glaciers of greater size in those regions of the coast with little local precipitation, typical icebergs of the more well-known Arctic types are formed in great numbers each year.

From these places, they are carried north and west along the coast to join the main pack, where, however, they are overshadowed and fade into relative insignificance beside the great tabular icebergs, or "snowbergs" as they have somewhat incorrectly been called. In actual numbers, it is probable that the variously shaped glacier bergs would equal or surpass their more majestic neighbours, but the latter, with their clear-cut contours and their (often) immense size, are incomparably more striking to the eye, and must be responsible for the annual removal of a considerably greater amount of ice. Individually, the true glacier berg has perhaps the finer appearance, but nothing can be more impressive than the sight of a fleet of tabular ice monsters, often many miles in length, and averaging fifty to a hundred feet in height, sailing majestically in a calm sea, churning up the pack as they move irresistibly northward in the grip of the Ross Sea current. ...

The area of the greatest of the tabular icebergs almost surpasses belief, but the existence of single bergs up to thirty miles in length, many hundreds of miles from land, is now well authenticated. The *Challenger* Expedition, Sir James Ross, the *Nimrod* and the *Terra Nova* Expeditions, to mention only four cases, all observed icebergs between twenty and thirty miles long, which needed several hours' sailing or steaming to clear. Icebergs a mile or more in length occur in hundreds, and indeed the breaking away of ice in the Antarctic clearly takes place on a scale quite unknown elsewhere in the world. Mixed with these immense four-square bergs are thousands upon thousands of smaller ones of all sizes, of several types, in every stage of dissolution. In size they vary from the great leviathans already mentioned, through the lesser tabular bergs, to the greatest of the true glacier bergs, and thence again past the product of the cascading glaciers, to the small ice-foot bergs, and finally the growlers, which represent the last clearly distinguishable products of land ice. Beyond this stage, the "bergy bits" and "brash" derived from the denudation of icebergs are in no sense distinguishable – except in their intimate internal structure – from the similar results of the disintegration of sea ice.

The age of the Ross Sea iceberg varies greatly according to the incidents of its life history. Even under what should at first sight appear to be the most favourable circumstances – for instance, when stranded on a shoal in high latitudes – dissolution proceeds at great speed during the summer months, and the disintegrating forces are often by no means idle during the winter months. Two quite distinct cases therefore require consideration in this connection. They are, respectively, the case of bergs detained in Antarctic waters and the more normal case of bergs which are carried straight into the pack and so north to the open sea.

Very many icebergs formed in high latitudes are likely to have some difficulty in reaching the main pack-belt, and this is probably more so in the case of the true glacier bergs than the tabular bergs, the ice of which has a less density. Striking examples of the contrast were seen at Cape Adare in 1911. The submarine extension of the beach on which Camp Ridley is built lay athwart the tidal current in and out of the bay. Daily the current on the ebb tide brought its load of floes and bergs close past the point. It was most instructive to see low glacier bergs stranding comparatively far out, and tall majestic tabular bergs sailing steadily past inside them. Not once but on several occasions did well-marked cases of this occur. Such a difference in draught must have a considerable effect in concentrating tabular bergs in the outer pack-belt and detaining icebergs of glacier type, until nearly the end of their career, in Antarctic waters. A good example of delay in the passage north of an iceberg was provided by part of Glacier Tongue, which stranded for several months at Cape Bernacchi before moving on elsewhere.

At every winter quarters established in the Ross Sea area there have been at least several icebergs stranded, usually of true glacier ice, so that the study of their disintegration in high latitudes has been made easy. The earliest stage in the dissolution of such an iceberg is usually the rapid rounding of its more angular contours by the comparatively warm waters of the autumn sea. This is accompanied by a certain amount of undermining and calving of relatively unstable portions, while lumps of sea ice are often thrown upon its top during this and the next stage while the sea is still open. As the winter temperatures set in, the iceberg is next covered by a thick coating of spray, which gives it a very characteristic appearance, and which may be so thick as materially to alter its equilibrium. This coating renders bergs whose history has included the "stranded" phase to be easily recognisable (Plate CCLXVI).

Plate CCLXVI: A Ross Sea iceberg stranded in winter shows a thick coating of sea spray. Priestley noted that such bergs' contours had been rounded "by the comparatively warm waters of the autumn sea".

This stage of accumulation ceases with the final freezing in of the sea ice, and the passing winter is accompanied by a steady growth about sea level, often with the formation of a well-marked tidal platform. At the same time, solution may be taking place towards the base of the iceberg and some ablation in the upper portion exposed to the air, though it should be noted that the latter is not usually sufficient to remove all the spray ice which has been added to the berg in the autumn.

Changes of temperature meanwhile cause the formation of cracks, which do not again cement and hence render the berg peculiarly vulnerable to the assaults of the waves in the following summer. In a tabular berg in which cubical jointing is well developed, the frost action may result in a complete crumbling of the upper portion of the berg. This is well seen ... where the avalanching of the debris on the sea ice is proof positive that very considerable disintegration took place during the three months between the freezing over of the bay in which the berg was stranded and the visit made by the sledging party which took the photograph.

The above summary, which is the result of the observation of many icebergs stranded near winter quarters, gives a good idea of the amount of destruction which can be achieved under these circumstances by the weathering agents during a single autumn and winter. Such a berg is usually sufficiently broken up to permit it to clear the shoal on which it has been stranded immediately the sea ice goes out in the beginning of the following summer. It then continues its journey to the pack as several much smaller fragments, and it is likely that a single season in the pack will carry these smaller bergs sufficiently far north to ensure their disintegration in the following summer. The life of an iceberg thus detained in the south is not likely to be more than three years – though, exceptionally, its detention south of the main ice-belt may be prolonged for another year or more.

The history of a tabular berg, or a glacier berg which reaches the main pack early in its first season, is for several reasons somewhat different. One characteristic of a broad belt of ice such as the ice pack is an absence of swell and waves, and this effectually limits the solvent action of sea water to the part of the berg at or below the surface, and also prevents any accumulation of sea spray such as has just been described. At the same time the more constant temperature of the air over the open sea reduces the cracking effect due to change of temperature. Similarly, the abrading action of the sea in the next summer is again diminished. The general result is seen in the more regular outline of bergs whose whole life has been spent in the ice pack. Such regular bergs are very common, and in their case disintegration is to a great extent postponed until they pass through the outer streams of the ice pack into the open water to the north. Once this last step has been taken, dissolution must be relatively rapid, so that the age of the majority of Antarctic icebergs must be determined by the length of time spent in the pack. What this time is we cannot know with any certainty. A favourable association of winds and deep currents might conceivably keep the bergs within the ice belt for several years. What knowledge we have of the currents and winds of the Ross Sea area, however, suggests that bergs would receive few setbacks to their steady passage north while in the pack.

One noticeable feature of all icebergs seen in the Ross Sea pack has been their freedom from accumulations of snow. At first sight this might suggest that their stay in the pack has been short, but this fact is not to be relied upon as evidence. Owing to the comparatively small size of the majority of the bergs and the strength of the winds in the pack zone, accumulation of snow on the top of bergs is not to be expected to any great extent. Accumulations of snow in the neighbourhood of bergs will, of course, be continually broken up and carried away from the berg by the constant change in relative position of the pack elements. Another factor likely to prevent accumulation of snow on bergs in the Ross Sea area, is the relatively small precipitation … .

The only apparent exception to the comparatively short life of Antarctic icebergs is the case of bergs frozen in soon after they were formed over shoal water and embedded in a sheet of fast ice of more than ordinary persistence. The life of a glacier berg under these conditions is only limited by that of the sheet of shelf ice of which it may conceivably come to form a part.

Such cases are exceptional, few examples of the inclusion of icebergs in such sheets having yet been recorded, though the *West-Eis* of von Drygalski appears to originate in some such way. If sheets of shelf ice are formed from sea ice in this manner, however – and this seems to be conclusively proved – included bergs will quite likely have taken part in their formation, and indeed may have formed the "piles" which were the main strengthening elements of the original sea-ice base on which the sheet has formed. Such bergs would become an integral part of the shelf ice and would behave more or less in conformity with its other elements. They would then take part in the outward flow which would be the consequence of accumulation of snow above.

> **In the more detailed account of the life history of an iceberg that follows, Priestley writes about some of the curious and distinctive features of a weathering berg.**

A pronounced feature of the weathering of tabular icebergs, whether by atmospheric agencies or by sea water, is the change in colour from the original dazzling white to an equally beautiful blue. This result may be brought about in different ways, but it is due in the main to internal change in the structure of the ice, which is usually dependent upon the presence of water. Constant washing by sea water, or the exposure to relatively warm temperatures in summer, are the principal causes. The net result is increase in the size of grain of the ice and decrease in the air content. The same process under different conditions has been noted and described as giving rise to blue bands in glaciers.

Another phenomenon frequently associated with the denudation of icebergs of all types is the formation of caves. Such caves may also be formed in several ways. The commonest method perhaps is by the widening of vertical planes of weakness such as cracks, and horizontal planes such as bedding planes, by a mixture of mechanical erosion and solution. Another method quite common in bergs stranded in high latitudes is by the closing of the top of a crevasse by a deposit of frozen spray. Most bergs seen in the pack have caves at or near the waterline. Cases have been observed of fish caught up in the ice of bergs. Such occurrences are proof positive that, since the formation of the berg, it has received some addition – probably from frozen sea water. During the passage through the pack in 1910, a fish was thrown out upon one floe by the sudden upturning of another under the impact of the nose of the ship. Many times, seaweed and sea animals have been observed to be thrown up on the ice foot in storms, and frozen to steep ice slopes before they had time to fall or be washed back. Such cases could, of course, only occur at low temperatures, but instances frequently arise when stranded bergs are washed with sea spray at a temperature 40° or more below zero Fahrenheit.

There are, of course, other ways in which fish might be trapped in water pools in icebergs and subsequently incorporated in the main mass of the berg through freezing. At Cape Adare it was quite common to see small fish playing about in the surface water near bergs, sometimes right within undercut caves whose floor was below sea level. A sudden change in equipoise, resulting in the tilting of the berg, would suffice to cut off a considerable quantity of sea water and trap the fish. Debenham's explanation of the mode of formation of the deposits of organic remains, so frequently found at the surface of glaciers, would also account for the presence of remains of fish in icebergs at the time of their separation from the parent mass.

Jointing and planes of stratification must often play a decisive part in assisting the quick weathering of icebergs. Cases have been met where the whole top ten or twenty feet of a tilted stratified berg has slid off bodily into the sea, the split taking place along a bedding plane where for some reason cementation has been less perfect (Plate CCLXXVII, [page 158]). The possibility of such a thing happening is suggested by the ease with which snow sometimes breaks along the junction between two successive drifts. The quick deposit of soft snowdrift upon a very hard polished surface must constitute a line of weakness in a stratified ice mass for many years after its consolidation. Unusually soft layers of snow must also at times act like layers of shale in rock complexes.

Other cases have been seen in which such a slide has not been quite complete, leaving cubical remnants standing up like the "buttes" or monadnocks standing out from an otherwise peneplained, horizontally bedded country. Such isolated remnants are easily mistaken for sea-ice blocks at a distance.

The effect of jointing and crevassing on the weathering of icebergs is well seen ... . Any such lines of weakness will provide an excellent point of entry for the tools of all the degrading forces. The rocking of the berg on a swell will tend to shake out blocks. The waves of the sea will find so much more surface on which to work, both by solution and mechanically. The innumerable shocks experienced by the berg in passing through close pack will have more planes along which they can take effect. The stranding of a much seraced, jointed, or crevassed berg on a shoal may well cause its utter ruin and dissolution into a number of smaller pieces. Instances have been seen of a berg running ashore at a speed of three or four knots, and of the seracs on its upper surface collapsing as the upperworks of a sailing ship have often done in similar circumstances.

Examples of sudden breaking up from apparently quite inadequate causes have been observed, but in such cases the cause must be looked for in the solution of the submerged portions of the berg, which may, under certain circumstances, go on much more quickly than weathering above the water level. One of the most dangerous features about an iceberg, when a near neighbour to a ship, is that nothing certain can be known about the state of the submerged portion, and therefore of the stability of the whole berg. Sudden submarine calving, with consequent overturning, is a well-known feature of all the later stages in the iceberg's history. Many of the nondescript blue-ice bergs seen owe their apparent anomalous shapes to complete overturning, a fact which can often be deduced from the disposition of the original stratification or dirt bands.

The geological work of icebergs during their short life is significant in several ways. As a distributor of rock, the iceberg is certainly of more importance than sea ice. The apparent virgin purity of the normal Antarctic iceberg is belied by the appearance of many of the tilted and overturned bergs in the later stage of decay. Such bergs are found frequently to contain considerable quantities of rock material. Even the apparently white tabular bergs, when reduced in size by thaw and ablation, often prove to contain a not inconsiderable amount of foreign matter. Good examples have been observed of icebergs stranded along the coast, from the upper layers of which dust and gravel have been concentrated until rock drifts an inch or more in thickness have been formed. This, and the heavier material carried lower down in the ice, are all dropped on the sea bottom, and at the present stage of Antarctic glacierisation must form no inconsiderable proportion of the bottom deposits. Much of the material dredged from the floor of Antarctic seas consists of rounded and ice-worn stones which must have been deposited in this way. A notable example, of particular interest because of its geological significance, is that of the haul of Cambrian limestone boulders brought up by the *Scotia*'s dredge off Coats Land. By far the greatest part of our knowledge of the Antarctic Cambrian fauna has come from an examination of these rocks.

Another less important work carried out by icebergs is that which results from their frequent stranding upon shoals. At certain points along the coast of South Victoria Land this is well shown, and particularly so at Cape Adare. Here the tide runs twice a day past the end of the Ridley Beach. Icebergs are continually stranding on and bumping over the shoal, and the bergs keep this portion of the sea bottom completely free from life, while they must considerably alter the contour of the bottom, both by gouging and by carrying off pebbles with them when they resume their journey at a later date.

The effect of bergs charging glaciers and sea ice is obvious. Both to the charged and the charger the result is disastrous, and such incidents are so frequent that they must play quite a noticeable part in preventing a greater accumulation of ice around the Antarctic continent.

Icebergs in pack move always at a different rate from the other constituents of the pack. Wind has less effect upon them, so that normally they move more slowly. The pack in rear of them therefore surges past under the urge of the wind, giving rise to the curious optical delusion that the bergs themselves are moving in the opposite direction. The presence of heaped-up pressure to windward and open lanes to leeward of a berg is a common phenomenon in the pack. The reverse process is seen when the wind changes and blows from a different quarter. Then the current in the lower layers of the sea still causes the bergs to move forward in the old direction, after the direction of movement of the surface water and the sea ice has changed. A similar contrary effect may be seen frequently when, as often happens, bergs are moving in a different direction from the pack ice owing to the presence of a deep-water current.

It is under such circumstances that they naturally perform the greatest amount of destructive work, and it is then that they are most dangerous to ships.

The later stages in the dissolution of icebergs, usually in the summer after their formation, is marked by the formation of smooth water-shaped contours ... . This stage is marked by the occasional complete submergence of the berg during storms, and by frequent changes of equilibrium. Before this stage is reached, however, it may happen that, in a tabular berg where the natural lines of weakness are most prominent, denudation may take place rapidly above water, while there still remains a broad stable pedestal below. In such cases, iceberg "swan ice" may be very well developed, as shown in Plate CCLXXXI [page 171]. An illustration of another peculiar form, etched by solution and complicated by the deposition of the frozen spray of a second autumn, is illustrated in Plate CCXVIII [page 172].

Such rounded contours as are found in all sea-worn bergs are in

Plate CCLXXXI: Priestley postulated that, in the summer months, tabular icebergs often wore away rapidly above water, while retaining a broad stable base under it. "In such cases," he wrote, "'swan ice' may be very well developed."

marked contrast to the sharp outlines common on the seaward faces of glaciers. This contrast, of course, is a result of the difference between the part played by water in the weathering of the former and of the latter. Even on bergs stranded in high latitudes, thaw-water has much more effect than on neighbouring glacier cliffs. This may partly be attributed to the fact that the surface of all such bergs is coated to some extent with salt ice. The brine produced from the thawing of this must greatly assist in its turn to hasten the thawing of the salt-free ice beneath. In a similar way, the first signs of melting about a winter quarters occur either in drifts strewn with rock fragments and full of dust, or in the salty ice of the ice foot and fast ice.

Plate CCXVIII: Swan ice at Cape Adare. Priestley noted: "Such rounded contours as are found in all sea-worn bergs are in marked contrast to the sharp outlines common on the seaward faces of glaciers."

The effect of surf and breaking waves in disintegrating bergs is much assisted by the presence of the caves formed earlier in their career. This could particularly well be seen where the caves had been so numerous as to occupy a relatively large proportion of the iceberg's waterline. In such cases, it was quite common for the surf to batter its way right through to the surface of the berg, with the formation of great blowholes through which spray was thrown many feet into the air. The next stage would be the breaking away of large pieces of the roof of the cave, with the formation of "bergy bits" and "growlers," which are the penultimate product of the disintegration of land ice at sea.

From "Ch XII: Antarctic Icebergs" by Raymond Priestley, in *Glaciology: British (Terra Nova) Antarctic Expedition 1910–1913* by C.S. Wright and R.E. Priestley: Harrison & Sons, London, 1922.

Richard Byrd's first Antarctic expedition built a sturdy base, "Little America", near the Bay of Whales, an indentation on the Ross Ice Shelf, in 1928. Here construction is underway.

Byrd makes a meteorological observation

After Amundsen and his party reached the South Pole and Scott and his party died on their return journey, there was a lull in Antarctic exploration and science as the First World War diverted resources away from exploration and natural history. United States expeditions between 1928 and 1940 brought new technologies – most notably telecommunications and motorised transport – to the continent. This account is by the American Admiral Richard Byrd, best known for flying solo to both the North and South Poles, in 1926 and 1929 respectively, although his North Pole claim was disputed. On his second trip to Antarctica in 1934, Byrd chose to spend the winter in a small, buried, purpose-built hut – "Advance Base" – at 80° South, making meteorological and auroral observations, while the rest of his team remained at the main base, "Little America", on the East Ross Ice Shelf, inland from the Bay of Whales. When there was radio contact the team at Little America could convey voice messages but Byrd could communicate only, and inexpertly, in morse code.

I was not long in discovering one thing: that if anything was eventually to regularise the rhythm by which I should live at Advance Base, it would not be the weather so much as the weather instruments themselves. I had eight in continuous operation. One was the register … which kept a continuous record of wind velocities and directions. The electrical circuit, connecting with the weathervane and wind cups on the anemometer pole topside, was powered by nine dry-cell batteries, and the brass drum with the recording sheet was turned by a clockwork mechanism which I had to wind daily. The sheet was lined at intervals corresponding to five minutes in time, and between these lines two pens, one representing the speed of the wind and the other its direction, wrote steadily from noon of one day to noon of the next.

Two other instruments were thermographs, which recorded temperature changes. The so-called inside thermograph was a fairly new invention, whose unique virtue was that it could be housed inside the shack. A metal tube filled with alcohol projected through the roof, and the expansions and contractions of the liquid in the tube drove a pen up and down over a rotating sheet set in a clock-faced dial hanging from the wall, just over the emergency radio set. The sheet, marked with twenty-four spokes for the hours and with concentric circles for the degrees of temperature, made one rotation in twenty-four hours; it would record accurately down to 85° below zero. The outside thermograph was a compact little mechanism which served the same function, except that it stood in the instrument shelter topside and the sheets needed changing only once a week.

Besides these instruments, I had a barograph to record atmospheric pressure, which was kept in a leather case in the food tunnel. Plus a hygrometer, employing a human hair, for measuring humidity (not very reliable, though, at cold temperatures). Plus a minimum thermometer, which measured the lowest temperature. In it was a tiny pin which was dropped by the contraction of alcohol in the column. Alcohol was used instead of mercury because mercury freezes at $-38°$ whereas pure grain alcohol will still flow at $-179°$. This instrument was useful as a check on the thermographs. It was kept in the instrument shelter, a boxlike structure set on four legs, which stood shoulder-high, close to the hatch. The sides were overlapping slats spaced an inch apart to allow air to circulate freely and yet keep out drift.

If I had had any illusions as to being master in my own house, they were soon dispelled. The instruments were masters, not I, and the fact that I knew none too much about them only intensified my humility.

There was scarcely an hour in the living day of which a part was not devoted to them or observations connected with them.

Every morning at eight o'clock sharp – and again at eight o'clock in the evening – I had to climb topside and note the minimum temperature reading, after which I would shake the thermometer hard to put the pin back into the fluid. Then, standing five minutes or so at the hatch, I would consult the sky, the horizon, and the Barrier, noting on a piece of scratch paper the percentage of cloudiness, the mistiness or clarity, the amount of drift, the direction and speed of the wind (a visual check on the register), and anything particularly interesting about the weather. All of these data were dutifully entered on Form No. 1083, U.S. Weather Bureau.

Every day, between twelve o'clock and one o'clock, I changed the recording sheets on the register and the inside thermograph. The pens and the pads supplying them always needed inking, and the thermograph clock had to be wound. Mondays I performed the same service for the outside thermograph and the barograph. ...

By May 17th, one month after the sun had sunk below the horizon, the noon twilight was dwindling to a mere chink in the darkness, lit by a cold reddish glow. Days when the wind brooded in the north or east, the Barrier became a vast stagnant shadow surrounded by swollen masses of clouds, one layer of darkness piled on top of the other. This was the polar night, the morbid countenance of the Ice Age. Nothing moved; nothing was visible. This was the soul of inertness. One could almost hear a distant creaking as if a great weight were settling.

Out of the deepening darkness came the cold. On May 19th, when I took the usual walk, the temperature was 65° below zero. For the first time the canvas boots failed to protect my feet. One heel was nipped, and I was forced to return to the hut and change to reindeer mukluks. That day I felt miserable; my body was racked by shooting pains – exactly as if I had been gassed. Very likely I was: in inspecting the ventilator pipes next morning I discovered that the intake pipe was completely clogged with rime and that the outlet pipe was two-thirds full.

Next day – Sunday the 20th – was the coldest yet. The minimum thermometer dropped to 72° below zero; the inside thermograph, which always read a bit lower than the instruments in the shelter, stood at –74°; and the thermograph in the shelter was stopped dead – the ink, though well laced with glycerine, and the lubricant were both frozen. So violently did the air in the fuel tank expand after the stove was lit that oil went shooting all over the place; to insulate the tank against similar temperature spreads I wrapped around it the rubber air cushion which by some lucky error had been included among my gear. In the glow of a flashlight the vapour rising from the stovepipe and the outlet ventilator looked like the discharge from two steam engines. My fingers agonised over the thermograph and I was hours putting it to rights. The fuel wouldn't flow from the drums; I had to take one inside and heat it near the stove. All day long I kept two primus stoves burning in the tunnel.

Sunday the 20th also brought a radio schedule; I had the devil's own time trying to meet it. The engine baulked for an hour; my fingers were so brittle and frostbitten from tinkering with the carburettor that when I actually made contact with Little America I could scarcely work the key. "Ask Haines come on," was my first request. While Hutcheson searched the tunnels of Little America for the senior meteorologist, I chatted briefly with Charlie Murphy. Little America claimed only –60°. "But we're moving the brass monkeys below," Charlie advised. "Seventy-one below here now," I said. "You can have it," was the closing comment from the north.

Then Bill Haines' merry voice sounded in the earphones. I explained the difficulty with the thermograph. "Same trouble we've had," Bill said. "It's probably due to frozen oil. I'd suggest you bring the instrument inside, and try soaking it in gasoline to cut whatever oil traces remain. Then rinse it in ether. As for the ink's freezing, you might try adding more glycerine."

Bill was in a jovial mood. "Look at me, Admiral," he boomed. "I never have any trouble with the instruments. The trick is in having an ambitious and docile assistant."

I really chuckled over that because I knew, from the first expedition, what Grimminger, the junior meteorologist, was going through: Bill, with his back to the fire and blandishment on his tongue, persuading the recruit that duty and the opportunity for self-improvement required him to go up into the blizzard to fix a baulky trace; Bill humming to himself in the warmth of a shack while the assistant in an open pit kept a theodolite trained on the sounding balloon soaring into the night, and stuttered into a telephone the different vernier readings from which Bill was calculating the velocities and directions of the upper air currents. That day I rather wished that I, too, had an assistant. He would have taken his turn on the anemometer pole, no mistake. The frost in the iron cleats went through the fur soles of the mukluks and froze the balls of my feet. My breath made little explosive sounds on the wind; my lungs, already sore, seemed to shrivel when I breathed.

Seldom had the aurora flamed more brilliantly. For hours the night danced to its frenetic excitement. And at times the sound of Barrier quakes was like that of heavy guns. My tongue was swollen and sore from drinking scalding hot tea, and the tip of my nose ached from frostbite. A big wind, I guessed, would come out of this still cold: it behoved me to look to my roof. I carried gallons of water topside, and poured it around the edges of the shack. It froze almost as soon as it hit. The ice was an armour-plating over the packed drift.

At midnight, when I clambered topside for an auroral "ob", a wild sense of suffocation came over me the instant I pushed my shoulders through the trapdoor. My lungs gasped but no air reached them.

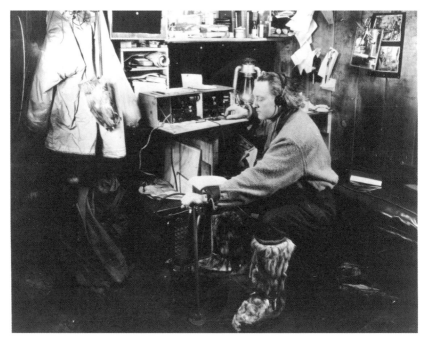

Richard Byrd taps out morse code in his small buried hut, "Advance Base", where he spent five months alone during the winter of 1934 in temperatures as low as minus 60°C.

Bewildered and perhaps a little frightened, I slid down the ladder and lunged into the shack. In the warm air the feeling passed as quickly as it had come. Curious but cautious, I again made my way up the ladder. And again the same thing happened; I lost my breath, but I perceived why. A light air was moving down from eastward and its bitter touch, when I faced into it, was constricting the breathing passages. So I turned my face away from it, breathing into my glove, and in that attitude finished the "ob". Before going below I made an interesting experiment. I put a thermometer on the snow, let it lie there awhile, and discovered that the temperature at the surface was actually 5° colder than at the level of the instrument shelter, four feet higher. Reading in the sleeping bag afterwards I froze one finger, although I shifted the book steadily from one hand to the other, slipping the unoccupied hand into the warmth of the bag.

Out of the cold and out of the east came the wind. It came on gradually, as if the sheer weight of the cold were almost too much to be moved. On the night of the 21st the barometer started down. The night was black as a thunderhead when I made my first trip topside, and a tension in the wind, a bulking of shadows in the night, indicated that a new storm centre was forming. Next morning, glad of an excuse to stay underground, I worked a long time on the Escape Tunnel by the light of a red candle standing in a snow recess. That day I pushed the emergency exit to a distance of twenty-two feet, the farthest it was ever to go.

My stint done, I sat down on a box, thinking how beautiful was the red of the candle, how white the rough-hewn snow. Soon I became aware of an increasing clatter of the anemometer cups. Realising that the wind was picking up, I went topside to make sure that everything was secured. It is a queer experience to watch a blizzard rise. First there is the wind, rising out of nowhere. Then the Barrier unwrenches itself from quietude, and the surface, which just before had seemed as hard and polished as metal, begins to run like a making sea.

Sometimes, if the wind strikes hard, the drift comes across the Barrier like a hurrying white cloud, tossed hundreds of feet in the air. Other times the growth is gradual. You become conscious of a general slithering movement on all sides. The air fills with tiny scraping and sliding and rustling sounds as the first loose crystals stir. In a little while they are moving as solidly as an incoming tide, which creams over the ankles, then surges to the waist, and finally is at the throat. I have walked in drift so thick as not to be able to see a foot ahead of me; yet, when I glanced up, I could see the stars shining through the thin layer just overhead.

Smoking tendrils were creeping up the anemometer pole when I finished my inspection. I hurriedly made the trapdoor fast, as a sailor might batten down a hatch, and knowing that my ship was well secured, I retired to the cabin to ride out the storm. It could not reach me, hidden deep in the Barrier crust; nevertheless the sounds came down. The gale sobbed in the ventilators, shook the stovepipe until I thought it would be jerked out by the roots, pounded the roof with sledgehammer blows. I could actually feel the suction effect through the pervious snow. A breeze flickered in the room and the tunnels. The candles wavered and went out. My only light was the feeble storm lantern.

Even so, I didn't have any idea how really bad it was until I went aloft for an observation. As I pushed back the trapdoor the drift met me like a moving wall. It was only a few steps from the ladder to the instrument shelter but it seemed more like a mile. The air came at me in snowy rushes; I breasted it as I might a heavy surf. No night had ever seemed so dark. The beam from the flashlight was choked in its throat; I could not see my hand before my face.

My windproofs were caked with drift by the time I got below. I had a vague feeling that something had changed while I was gone, but what I couldn't tell. Presently I noticed that the shack was appreciably colder. Raising the stove lid I was surprised to find that the fire was out, though the tank was half full. I decided that I must have turned off the valve unconsciously before going aloft, but when I put a match to the burner the draught down the pipe blew out the flame. The wind, then, must have killed the fire. I got it going again, and watched it carefully.

The blizzard vaulted to gale force. Above the roar the deep taut thrumming note of the radio antenna and the anemometer guy wires reminded me of wind in a ship's rigging. The wind direction trace turned scratchy on the sheet; no doubt drift had short-circuited the electric contacts, I decided. Realising that it was hopeless to attempt to try to keep them clear, I let the instrument be. There were other ways of getting the wind direction. I tied a handkerchief to a bamboo pole and ran it through the outlet ventilator; with a flashlight I could tell which way the cloth was whipped. I did this at hourly intervals, noting any change of direction on the sheet. But by two o'clock in the morning I had had enough of this periscope sighting. If I expected to sleep and at the same time maintain the continuity of the records, I had no choice but to clean the contact points.

The wind was blowing hard then. The Barrier shook from the concussions overhead, and the noise was as if the entire physical world were tearing itself to pieces. I could scarcely heave the trapdoor open. The instant it came clear I was plunged into a blinding smother. I came out crawling, clinging to the handle of the door until I made sure of my bearings. Then I let the door fall shut, not wanting the tunnel filled with drift. To see was impossible. Millions of tiny pellets exploded in my eyes, stinging like BB shot. It was even hard to breathe because snow instantly clogged the mouth and nostrils. I made my way toward the anemometer pole on hands and knees, scared that I might be bowled off my feet if I stood erect; one false step and I should be lost forever.

I found the pole all right, but not until my head collided with a cleat. I managed to climb it too, though ten million ghosts were tearing at me, ramming their thumbs into my eyes. But the errand was useless. Drift as thick as this would mess up the contact points as quickly as they were cleared; besides, the wind cups were spinning so fast that I stood a good chance of losing a couple of fingers in the process. Coming down the pole, I had a sense of being whirled violently through the air, with no control over my movements. The trapdoor was completely buried when I found it again, after scraping around for some time with my mittens. I pulled at the handle, first with one hand, then with both. It did not give. It's a tight fit anyway, I mumbled to myself. The drift has probably wedged the corners. Standing astride the hatch, I braced myself and heaved with all my strength. I might just as well have tried hoisting the Barrier.

Panic took me then, I must confess. Reason fled. I clawed at the three-foot square of timber like a madman. I beat on it with my fists, trying to shake the snow loose; and when that did no good I lay flat on my belly and pulled until my hands went weak from cold and weariness. Then I crooked my elbow, put my face down, and said over and over again, You damn fool, you damn fool. Here for weeks I had been defending myself against the danger of being penned inside the shack; instead I was now locked out. And nothing could be worse, especially since I had only a wool parka and pants under my windproofs. Just two feet below was sanctuary – warmth, food, tools, all the means of survival. All these things were an arm's length away but I was powerless to reach them.

There is something extravagantly insensate about an Antarctic blizzard at night. Its vindictiveness cannot be measured on an anemometer sheet. It is more than just wind: it is a solid wall of snow moving at gale force, pounding like surf. The whole malevolent rush is concentrated upon you as upon a personal enemy. In the senseless explosion of sound you are reduced to a crawling thing on the margin of a disintegrating world; you can't see, you can't hear, you can hardly move. The lungs gasp after the air sucked out of them, and the brain is shaken. Nothing in the world will so quickly isolate a man.

Half-frozen, I stabbed towards one of the ventilators a few feet away. My mittens touched something round and cold. Cupping it in my hands, I pulled myself up. This was the outlet ventilator. Just why I don't know – but instinct made me kneel and press my face against the opening. Nothing in the room was visible, but a dim patch of light illuminated the floor and warmth rose up to my face. That steadied me.

Still kneeling, I turned my back to the blizzard and considered what might be done. I thought of breaking in the windows in the roof, but they lay two feet down in hard crust, and were reinforced with wire besides. If I only had something to dig with, I could break the crust and stamp the windows in with my feet. The pipe cupped between my hands supplied the first inspiration: maybe I could use that to dig with. It, too, was wedged tight. I pulled until my arms ached, without budging it. I had lost all track of time, and the despairing thought came to me that I was lost in a task without an end. Then I remembered the shovel. A week before, after levelling drift from the last light blow, I had stabbed a shovel handle up in the crust somewhere to leeward. That shovel would save me. But how to find it in the avalanche of the blizzard?

I lay down and stretched out full length. Still holding the pipe, I thrashed around with my feet but pummelled only empty air. Then I worked back to the hatch. The hard edges at the opening provided another grip, and again I stretched out and kicked. Again, no luck. I dared not let go until I had something else familiar to cling to. My foot came up against the other ventilator pipe. I edged back to that, and from the new anchorage repeated the manoeuvre. This time my ankle struck something hard. When I felt it and recognised the handle, I wanted to caress it.

Embracing this thrice-blessed tool, I inched back to the trapdoor. The handle of the shovel was just small enough to pass under the little wooden bridge which served as a grip. I got both hands on the shovel and tried to wrench the door up; my strength was not enough, however. So I lay down flat on my belly and worked my shoulders under the shovel. Then I heaved, the door sprang open, and I rolled down the shaft. When I tumbled into the light and warmth of the room I kept thinking, How wonderful, how perfectly wonderful.

From *Alone* by Richard E. Byrd: Island Press, London, 2003.

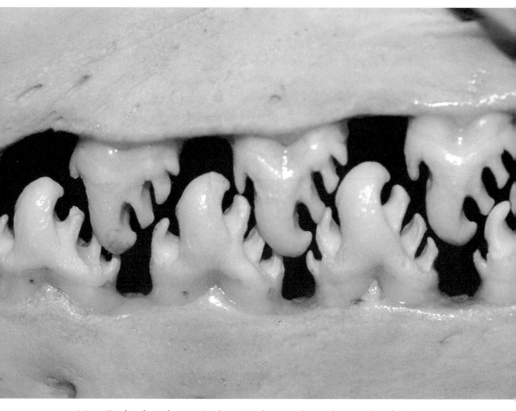

New Zealand zoologist Graham Turbott undertook a study of seals in the deep south and noted that only four were "truly Antarctic". These teeth are from the crabeater, which lives almost entirely on krill, filtering the tiny crustaceans through its teeth, which have permanent gaps to allow water to flow back out.

Crabeaters and leopards

Antarctic seals were of interest to hunters: in the 18th and 19th centuries some species were taken to the brink of extinction, long before much was understood about the animals' biology and ecology. In 1952 the New Zealand zoologist Graham Turbott noted that seals "are so distinctive and their adaption to aquatic life so striking it may be difficult to realise that so little is yet known of most species".

Turbott's first-hand knowledge of Southern Ocean seals came from his field trips to the subantarctic islands. In this extract from a book published by the New Zealand Antarctic Society, Turbott draws on research by other scientists, including British zoologists Edward Wilson, Colin Bertram and J.E. Hamilton, and Norwegian zoologist Nicoli Hanson, to give his accounts of the four "truly Antarctic" seals.

The Weddell seal is of heavy build and attains nine feet in length. The fresh coat is dark grey or almost black, with a handsome mottling of white and silver grey on the underparts. By the annual moult in late summer its colour has faded to a rusty grey.

Characteristic of the Antarctic coastline is the sight of the stout, slug-like Weddells lying in numbers scattered over the ice where they come out to sleep for long periods.

The strictly coastal habitat of this species has brought it immunity from the highly predatory killer whale. These seals also spend the winter in the inshore waters beneath the ice and depend upon breathing holes, which they must keep open by means of constant scraping with their large canine teeth. The usual procedure is for the seal to swing the head in a semicircle so the teeth saw through the ice, or the upper jaw may be rotated with the lower providing a fixed point. As a result, the "ice-sawing" teeth – the upper and lower canines and outer pair of upper incisors – generally show marked signs of wear or breakage in older seals. Many such handicapped animals probably die in winter through failure to open up the breathing holes.

Although the incisors and canines are comparatively strongly developed in the Weddell seal, the teeth are not specially adapted to a particular food, as in the case of all three following species. The food consists of both fish and cephalopods (squid), the fish being mainly of the Notothenia type so common in the Antarctic and Subantarctic.

In winter the air above is so much colder that the seals rarely leave the water except on very calm days. Much remains to be learned of the diving capacity of seals, but there is no doubt that Weddell seals wander far beneath the ice before returning to their breathing holes. It is remarkable that this is performed in the period when there is total darkness above, so navigation can receive no aid from even dim light which might penetrate through the ice. It is suggested that pockets of air beneath the ice may be used by the seals, but the critical factor in their existence is evidently the keeping open of the breathing holes.

Bertram gives a comprehensive account of the breeding cycle and the early growth of the young. There is a ten months' gestation, and the females collect on the ice in spring to bear the pups. Pupping is over within a month but the season is variable in different parts of Antarctica. The peak of pupping occurs in the third week of October in the Ross Sea, and just a month earlier in Graham Land. The sexes are approximately equal in numbers. Mating takes place in November and December but has never been seen, although it certainly occurs in the water soon after separation from the pups.

Weddell seal pups, which weigh sixty pounds at birth, grow at an astonishing rate, their weight being doubled in a fortnight. They have a thick woolly coat but a complete moult begins after two weeks and at the age of six weeks they have assumed the first coat of hair. The pup may enter the water at three weeks, but continues to be suckled until mother and pup separate permanently seven weeks after birth. Very rarely twins have been found, but the normal number of pups is one.

The young Weddell seal, as in true seals in general, is born at an advanced stage. It begins to obtain crustaceans for itself before the period of suckling is over, and for the first few months after weaning the pup continues to live on the same diet, which is gradually replaced by fish and squid.

In the first year the young grow to a length of about six feet, but in their second year rapidly reach adult size, and breed in the third year. The young rarely come ashore and little is known of their movements. It is thus interesting that several Weddell seals which have strayed to the subantarctic islands and to the Australian and New Zealand coasts have proved to be young animals not yet twelve months old.

The crabeater seal is the species seen first on approaching Antarctica. It is the seal of the pack ice, just as the Weddell is the species so characteristic of the Antarctic coast.

This species appears inshore in large numbers in summer, when the coastal waters around much of the continent are relatively free of ice. The influx has been observed on the coasts of the Ross Sea and western Graham Land, and according to the observations of Bertram there is a strong suggestion in the regularity with which the arrival and departure take place that – at least partially – the species is migratory. Details of the proportion which do not migrate and their distribution in summer are still unknown. There is also relatively little knowledge of the life of this species in the pack ice from the beginning of winter, when all have normally left the inshore waters, until the following summer.

The crabeater is smaller and more active than the Weddell seal, the maximum length being eight feet. The moult, which occurs in autumn, results in a remarkable change in colour, the fresh coat having previously faded to yellowish, and finally almost pure white. Upon moulting, crabeaters are a rich grey-brown, with irregular pale blotches, especially on the sides and lower back, giving a handsome dappled effect. But at the stage of greatest bleaching the crabeater is thoroughly deserving of the alternative name white Antarctic seal.

Besides being a much slimmer seal in general build, this species differs from the Weddell in its reactions when disturbed. Wilson states that "in attacking man or dogs, it rushes forward first with open mouth and husky roar, and then as quickly makes off for the nearest hole to reach the water". The appearance of the head is distinctive owing to the length of the snout, which, quoting again from Wilson, is "distinctly pig-like, and can be given a turned-up, truncated look when the animal is in fear or otherwise excited".

It is the most gregarious of the Antarctic species, commonly herding together in groups on the pack ice, and even in inshore waters prefers to crowd on to small floes, and only rarely comes ashore. It is rapid and agile on snow and ice and has developed a characteristic mode of progression, of which Wilson writes: "Its movements when alarmed become as nearly as possible quadrupedal, for while the lithe and active body takes on the motion of a fish, the fore limbs, instead of lying idle along the sides as in Weddell's seal, assume an alternating action, exactly as they would in a four-legged beast."

The food of the crabeater consists almost entirely of the shrimp-like euphausiid crustaceans so plentiful in the Antarctic and known as whale feed or krill. The cheek teeth are adapted as a sieve, each tooth being divided into several recurved lobes; a series of openings is thus left when the jaws are closed, by means of which the water is strained out of the mouth. Small fish are occasionally taken, but crustaceans alone have been found in most of the specimens examined.

The birth of the pups, which occurs in early spring, coincides with the time of greatest extent of the pack ice. As a result there is practically no information on the early life of the pups up to the age of three months, although they were seen in the pack ice during the drift voyages of the *Belgica* and the *Endurance*. Observations made by these expeditions indicate that the pups remain on the ice for a period of probably only three or four days after birth, and the first coat is lost within a fortnight.

Bertram, on the British Graham Land Expedition 1934–37, carried out research on the ovaries and development of the embryo of this species, and showed that the period of gestation is approximately nine months. The pups are probably born in September. It was found that the young crabeater has a much more rapid growth to maturity than the young Weddell seal, and attains breeding age in its second year.

From the position and nature of the open gashes and scars which have been seen so often on crabeaters, it seems probable this seal is one of the most frequent victims of attack by killer whales.

Crabeaters stray to the north only rarely, but stragglers have been recorded from South America, Australia and New Zealand.

The Ross seal, unless some stronghold awaits discovery, must be the rarest of all seals. According to Bertram less than fifty have been seen since the species was discovered.

Its food is chiefly cephalopods – cuttlefish and squid – while more rarely it may take crustaceans and fish. The canine and incisor teeth alone are fully developed, their curved, needle-like form being an adaptation enabling the prey to be seized.

Apparently, like the crabeater, the Ross seal is truly an inhabitant of the pack ice, but of quite distinctive diet and general characteristics. It has almost always been seen singly, and seems to be of solitary habits. On the ice it is lethargic, according to some accounts, although Hanson described a Ross seal which, when pursued, "made off again at a great speed".

A remarkable aspect of the appearance of this seal is its swollen blubbery neck, into which the head may almost disappear when withdrawn. Both fore and hind flippers are comparatively broad, suggesting it may surpass other Antarctic seals in its swimming powers.

In length the Ross seal reaches about nine feet. It is greyish-brown or black above and paler below, with a distinctive pattern of oblique pale streaks on the sides.

The leopard seal is distributed from the Antarctic coast and pack ice into subantarctic and even temperate regions. The three species of seal described above are restricted in range to well-marked regions within the Antarctic, each being limited to certain littoral or pelagic foods. The wider distribution of the leopard seal is clearly in accord with its more varied food habits.

The leopard seal is by nature a hermit, and individuals of this species probably wander extensively over the Southern Ocean. Whether in the pack ice or on the shores of subantarctic islands its occurrence is scattered, and its indifference to others of its species when they chance to be in the same locality is most marked.

Its diet includes fish and cephalopods, but the bulk of its food is obtained by preying upon other seals and especially seabirds. Probably the most important single food is penguins, both Antarctic and subantarctic. When a cormorant, petrel or penguin is killed, the body is brought to the surface and shaken so vigorously that it is at least partially skinned.

The carnivorous habits of the leopard seal are carried to the extent of feeding on the carcasses of whales and seals. There is no doubt that on occasions it kills adult seals of other species, and the younger animals and pups when opportunity offers. Hamilton describes an incident in the pack when a leopard seal jumped on to the ice and killed and partly consumed a crabeater seal. Such aggressively carnivorous habits are not found in any other species, although some eared seals, especially the fur seals of the south, regularly feed upon penguins and other seabirds.

The leopard seal is undoubtedly a fearsome beast of prey, with respect both to its size and the formidable development of its teeth, which form an array leaving no doubt as to its ability to attack. These include, in addition to the large curved canines and incisors, a row of triple-cusped, blade-like cheek teeth.

In young leopard seals the combination of slim tapering body and comparatively large head is particularly characteristic, and roughly the same shape is retained throughout life by the male. The proportions of the female change, the fore part of the body becoming much broader.

Leopard seals grow comparatively slowly until breeding age is reached in the third year in the case of the female and in the fourth year in the male.

The colour is dark grey, shading to paler grey below, and the neck, shoulders and flanks are distinctly splashed and spotted with silver and black. In immature animals the coat is a more silvery grey than in the adults.

Little is yet known of the breeding cycle of this seal, but the breeding season seems to extend for three months or more, approximately from September to December. The period of lactation is probably only a few days.

There is possibly a fairly general migration of leopard seals towards the northern part of their range in winter. According to Hamilton, this species is most plentiful on the Falkland Islands in spring and early summer and at Macquarie [Island] from late winter to early summer.

From "Chapter 7: Seals of the Southern Ocean" by E.G. Turbott, in *The Antarctic Today: A Mid-century Survey by the New Zealand Antarctic Society* edited by Frank A. Simpson: A.H. and A.W. Reed, Auckland, 1952.

Continent for Science

After the burst of activity that took place during the first two decades of the twentieth century, there was a lull in Antarctic science. The poles had been reached and people had died along the way. The First and Second World Wars focused resources elsewhere, while at the same time leading to massive technological advances and new scientific knowledge.

The next wave of Antarctic science would be a lot more modern and familiar to us today than the rugged and difficult science of the heroic age. It began during International Geophysical Year, IGY, which ran from July 1957 to December 1958. IGY Antarctic programmes focused on geophysical sciences, such as the study of cosmic rays, the aurora, geomagnetism, gravity, meteorology, oceanography, glaciology, seismology and solar activity.

Twelve nations – Argentina, Australia, Belgium, Chile, France, Japan, New Zealand, Norway, South Africa, USSR, UK and US – took part. By the end of IGY there were forty-eight bases in Antarctica. Most were on islands or around the coast of the continent but some were established high on the Polar Plateau, among them the Amundsen-Scott South Pole Station and a temporary Soviet base at the Pole of Inaccessibility, so named because it is the farthest point from the Southern Ocean. But one of the most significant things to come out of IGY was the Antarctic Treaty, which came into force in 1961. This banned military activities on the continent and defined Antarctica as a "continent for science".

Antarctic science expanded and diversified, as new technologies and scientific advances allowed scientists – and from the 1960s this included female scientists – to explore new parts of Antarctica. People dived under the sea ice, discovered extremophiles living in glacier ice and volcanic vents, and mapped the land that lay beneath ice sheets up to three kilometres thick. Public money and scientific collaboration allowed the funding of massive international projects to study the Antarctic marine ecosystem, drill sedimentary and ice cores that could reveal Antarctica's past climate, and embark on sophisticated astrophysics projects such as the Ice Cube Neutrino Observatory at the South Pole.

By the early twenty-first century, Antarctica had some of the world's most technologically sophisticated buildings, such as Britain's Halley VI, a modular station designed to stay above the snow, and away from the ice edge, by moving on ski-equipped hydraulic legs. Scientists working outside wore layers of high-tech insulating, windproof and waterproof clothes and boots designed to keep them functioning down to minus 100°C. They travelled by ski-plane, helicopter or snowmobile to field camps, where they communicated using satellite phones and wi-fi powered by diesel generators, solar or wind power. During the sunny summer nights, however, many campers still slept in Scott polar tents featuring the pyramid-based design used by Robert Falcon Scott on his 1911–12 campaign for the pole.

Physicist Colin Bull surveying from 1,823-metre "Peak Alpha" in the McMurdo Dry Valleys of Victoria Land.

Innocents in the Dry Valleys

In the summer of 1958–59 English-born physicist Colin Bull, who worked at Victoria University in New Zealand, led a two-month expedition to the McMurdo Dry Valleys that lie in Antarctica's Victoria Land. His mission was to establish survey stations, make gravity measurements to help determine the tectonic make-up of the valleys, and collect igneous rocks. Paleomagnetic data from these rocks would be used to help establish the location of Antarctica through geological time.

On the expedition with Bull were a biologist, Dick Barwick, and two undergraduate geology students, Barrie McKelvey and Peter Webb. All three had visited the Dry Valleys the season before. For most of the expedition Barwick worked as Bull's field assistant on the surveying work, investigating mummified seals and pond life when he found them, while McKelvey and Webb mapped the geology. As fellow Antarctic scientist Terence J. Hughes noted in his review of *Innocents in the Dry Valleys*, the book Bull wrote about the trip: "Bull believed that any one scientific discipline could be understood best if it were placed within a context of knowledge obtained from all other disciplines."

In this passage the four men have been flown by helicopter from Ross Island to their campsite in the Wright Valley, which is still largely unexplored. Bull begins by describing the view from his tent.

In the immediate foreground, twenty metres away, was the meteorological screen, erected on its pile of granite slabs and now equipped with barograph, thermograph, hygrograph, a handheld anemometer and sundry thermometers. Close by were the food boxes, with Maurice's frozen blocks of chicken stew and cans of tinned fruit piled up at the side. We had covered the food with a small tarpaulin sheet at first, not that we were expecting rain. A kilometre away was the ice-covered lake, with a freshwater moat initially a few metres wide at its edge but later in the summer maybe ten metres wide in most places. The lake was more than five kilometres long. On the north side, where the lake abutted the steep scree slopes, several benches were clearly cut by former higher stands of the lake. We wondered how long ago that had been.

Beyond the western end of the lake the valley divided into two, separated by a flat dolerite-topped mesa, and those we named straight away as the North and the South Forks. Obviously the valley floor rose to the west in the forks, but from the tents we couldn't see that area. But beyond that were two thin icefalls carrying a relatively small amount of ice from inland. At the bottoms of the falls the ice coalesced to form the apron glacier, just a few kilometres long, that we had seen on our recce flight.

Clearly lots of these features needed names but for the present we would name only the most pertinent. The glacier at the top end of the valley we called the Upper Wright Glacier, to match the Lower Wright Glacier at the other end. When bureaucracy ordained that they be called the Wright Upper Glacier and the Wright Lower Glacier we didn't object too much. How could we?

At Peter's suggestion we christened the flat-topped mesa with the name Dais, French for platform. The previous year he had called a similar flat-topped feature, separating two valleys in the Victoria Valley system, Insel, German for island.

Now for the lake! Peter had called the two prominent lakes in Victoria Valley Vida and Vashka, so it was quite apparent we needed another "V" lake and thus it became Lake Vanda, named after the lead dog in the only dog team I had ever driven seriously – in Greenland during the dark of winter of 1952–53. ... At that point we had exhausted our imaginative powers, so for the moment the peaks on the valley sides were named A, B, C, D … and even I knew that the next one along the range was E, as Guyon had done on his sketch map.[1]

In the other direction, eastwards towards McMurdo Sound, far in the distance we could clearly see Mount Erebus, 150 kilometres away or more. Constantly we were amazed at the clarity of the atmosphere. Barrie said, "Down here you can see forever, but it's even further than that to walk." From the tents we couldn't see the Wilson Piedmont Glacier or its offshoot the Lower Wright, or Wright Lower, Glacier, because of the bend in the valley. However, we could see the cliff-sided hanging glaciers, extending halfway to the valley floor on the south side and, somewhat imaginatively I thought, we called them Glaciers 1, 2, 3, 4, and – go on, guess! – 5, and later after professors at Vic, not knowing they had already been given names.

Bob Nichols, a geology professor at Tufts College, Massachusetts, and a great old Antarctic explorer – he held some record for the longest unsupported dog-sledge journey, with Ronne's expedition in 1947 – had been working at the Marble Point airstrip site in 1957 with a bunch of his students. One day they crossed the Wilson Piedmont Glacier (I don't know how) to the end of Wright Valley, which Bob called "The Grand Canyon of Antarctica" and named the hanging glaciers for his students: Denton, Goodspeed, Hart, Meserve and Bartley.

Our first full day in the valley started when the alarm woke me at 7.45 a.m. for a radio schedule with Peter Yeates, the friendly radio man at Scott Base, call sign ZLQ. Both the official radio, which belonged originally to the Trans-Antarctic Expedition, and the ship-to-shore ones that we had borrowed from the makers worked well – surprisingly well for the little ones, considering the distance. We sent them a message thanking them all for their tolerance and helpfulness. We meant it too, but Peter Yeates did seem to appreciate our saying it. After our

---

1 Guyon Warren, along with fellow geologist Bernie Gunn, had pioneered the exploration of the valleys the previous year.

conversation Peter "patched" me into WWV, the radio station that transmits time signals twenty-four hours a day. I needed to know the time within a second in order to calculate our position from observations of the sun. ... Both Dick and I carried "deck watches" in leather pouches around our necks to keep them at a constant temperature. They worked amazingly well as long as we remembered to wind them.

We found the charts for the recording met instruments, and, after adding the ink to their pens, my first task was to take dozens of sun sights with the theodolite throughout the day, to work out our position. It was warm, 4°C, but still a bit windy at the exposed instrument. Dick was luckier: as my "booker" (for angles and times of observations) he built an armchair in the rocks and sat there, out of the wind, chewing his pipe, and reading a book when I had to stop to adjust the levelling of the theodolite. I then tried setting up the subtense bar to measure the baseline, but the bar was shaking too much in the wind.

Meanwhile, Peter and Barrie had been sorting out the gear for their first exploration, which was to be a traverse of the range above the south side of the valley, starting from the most imposing peak visible from the valley floor, which was marked on our maps as Peak 105 and which we later called Mount Odin. Eventually they set off, with thirty-five-kilogram packs, at five p.m. At eight p.m., on the radio, they said that the scree was so difficult that they had to relay their loads. Well, if those superbly fit twenty-year-olds were having problems, how would Dick, aged twenty-nine, and I, overweight and all of thirty, manage? I looked again at Dais with its wonderful central position and flat terrain but accessible, as far as we could see, only by steep scree slopes followed by vertical dolerite rock bluffs, thought about the weight of the tripod and theodolite, and put a line – for the time being, at least – through Dais on our list of planned survey sites.

Wearing his four-point instep crampons, Dick walked across the rough surface of the lake to build a cairn and plant the inevitable

Day-Glo flag near the end of the peninsula that extended more than half of the distance across the lake. He tried to walk back without crampons, fell heavily, decided to walk round the lake edge, and found a mummified seal. Back at the tent he announced, "No fall without profit!" and extolled the beauty of his wretched seal.

I'd just gone to sleep when, at two a.m., the crunch of boots on granite shards woke me up. Peter and Barrie had taken five hours to carry half of their loads up to 1,300 metres, to an old glacier cirque between two peaks that rose another 600 metres above them. We had selected one of those peaks as a possible survey point. I wondered if we should delete that one as well. Peter said that carrying the packs up those steep scree slopes in near gale conditions was pretty close to purgatory. They came back down in two hours, but rather than face carrying the other half of their load up to the top, they walked back to camp with just their sleeping bags. Barrie was particularly effervescent and equally full of pains and puns – even after that gruelling trudge. Peter seemed to be most concerned with what Bob Clark [professor of geology at Victoria University of Wellington] would think if he knew his precious Exacta telephoto lens was buried in sand and snow at 1,300 metres. Ah, youth! We fed 'em and went back to bed.

Everywhere we looked there were interesting questions. Take the lake for example – ah yes, Lake Vanda. A stream flowed inland all the length of the valley from the Lower Wright Glacier, not a big stream: just before it flowed into the lake it was confined to a channel about three metres wide, and there it was thirty centimetres or so deep. There were a few other, and much smaller, streams flowing in to the south side of the lake, but I bet there was no outflow at the west end of the lake, because we could see the land rising beyond the end of the lake. Yet the water level had gone down about a couple of centimetres on the bamboo marker I'd stuck in the lake edge at noon the day before. Evaporation?

Then, why wasn't there any moraine in this part of the valley? Over there in the South Fork there seemed to be something that looked like a lobe of moraine. Or maybe it was a rock glacier. But there was nothing at all in this part of the valley. Could fluvio-glacial outwash from some larger ancestral Lake Vanda have flushed the valley clear? ...

The wind in the morning was cold and was from the east! Why was the wind from the coast colder, much colder, than the wind from the inland ice? Shouldn't it be the other way round? Nearly always the wind started to blow from the east at about five p.m., changing back again to blow from the inland ice at perhaps one a.m. With the easterly winds the humidity rose to about eighty percent, while the winds from the west were much drier, forty percent, as well as being warmer. But whatever its direction the wind was usually strong and sand-laden, and we had sand in our diaries, our eyes, our food, the theodolite, and our noses – everywhere. No wonder the place was covered with sand-eroded rock.

It was too windy to survey the baseline, so we inspected the mummified crabeater seal that Dick had found partially buried in the sand just forty metres from the tents. It was quite small and therefore young, Dick argued. The upper part was eroded down to the bone, but there was still fur on the underside. This was the first of ninety-nine seals, nearly all crabeater seals and nearly all young, that we found in the area. We wished we knew enough seal psychology to understand why they ventured so far from their natural habitat and into such an inhospitable place. Anyway, Dick photographed it, measured it (163 centimetres long), looked at its most recent dinner (granite sand) and wrote it all down, most assiduously, in his Seal Log Book.

Supper was another of Maurice's frozen chicken stews, which were slowly melting despite our best attempts at insulation. After supper the wind dropped, so I started work on setting up and measuring a baseline for our survey work. Originally I was going to put the

Dick Barwick, biologist on Colin Bull's New Zealand expedition of
1958–59, photographs a mummified crabeater seal in the Wright Valley.
It is not known why seals sometimes head inland to certain death.

markers for the ends of the baseline on the lake ice because that was the flattest area around. But then it occurred to me that, since the ice was floating, it might move in a heavy wind, so I put the markers on readily identifiable points on the southern shore of the lake, one where Dick had already built the cairn on the peninsula that extends most of the way across the lake.

Doing my own recording of angles was tedious but Dick wanted to do a first inspection of likely habitats for things that grow, whether plant or animal. Many of the local ponds, in the depressions between the dykes, were partly or completely thawed. In one of them, fully ten centimetres deep, Dick found the water temperature was 3°C and, to his great joy he found some creepy crawlies, which, he said, were called nematodes. A bit later when, dressed in a U.S. parka belonging to "POE", he had come down to the lake edge to see how I was doing with the survey work, he found some other things that moved, which he pronounced to be rotifers. He was practically ecstatic. He was equally joyful when after dinner we heard odd noises, which turned out to be a pair of skuas, disputing a discarded remnant down at the heliport. Apart from the nematodes and bits of algae and lichen, they were the first living things, other than our contubernial companions, that we had seen in the valley.

Ten days before Christmas, Dick and I set off to tackle our first surveying chores, hoping to establish three or four stations on the peaks on the north side of the valley before returning to the base for Christmas. Food for ten days, cooking gear, first-aid kit, tent, Dick's minimal collecting kit, sleeping bags, radio, Wild T2 theodolite and tripod, cameras, film, down jackets, gloves, alarm clock, climbing rope, notebooks, all came to something ridiculous, more than thirty-six kilograms each. Fortunately we had excellent rucksacks, items called Mountain Mules, one of the results of my letter solicitations back in Wellington. One of their best features was the hollow tubing of their frames, which could be filled with kerosene. Dick filled his with Drambuie.

I chucked out of the load all of my spare clothing and one of the two books, leaving me with *Gallant Gentlemen, A Portrait of the British Officer, 1600–1956* by E.S. Turner. Dick took John Masefield's *William Shakespeare,* a much better choice. When he tried lifting his loaded pack he poured the Drambuie, most of it anyway, back into the bottle.

Barrie McKelvey and Peter Webb also headed away from the main camp to map the valley's geology and take photographs from any peak that offered a good view.

Dick and I arranged future radio schedules with Peter and Barrie, and then spent the rest of the morning sorting out the stuff for our own trip. After lunch, in a tiresome brisk wind we set off for the col leading from Wright Valley to the valley later named McKelvey, part of the Victoria Valley system to the north. As we walked across the valley floor, up and down across the swarm of raised dykes in a rising wind, I regretted having lost the toss so that I was carrying the tent in its bag, more than a metre long, across the top of my pack and was being blown every which way. Dick had the tripod on top of his pack, also more than a metre long, but the wind could blow through that! It was obvious that I was being discriminated against! When we set out I thought of lots of pleasant things: Paremata and family, our canoe on Paremata Harbour and the like. Shortly I was thinking of John at Scott Base and his comment when changing reels on the projector: "Only another ten thousand feet of this excruciating garbage." That applied directly to us now. Our first survey station, marked "Peak 16" on the best photograph we had, was "only" about 1,900 metres above us. Obviously it was impossible to carry my load that high and that far, but I felt that with luck I could stagger to that big rock fifty metres away, couldn't I? And, with a little rest every few minutes, we continued.

Pecten shell fossils found by the group on Bull Pass. Peter Webb and Dick Barwick were right: these are now taken as evidence that pectens lived in Wright Fiord in the early Pliocene.

Heading towards the pass we moved to the north side of the valley, where we soon met the slopes of ancient, much subdued moraines. Over thousands of years the relentless wind had concentrated sheets of sand in the depressions between bouldery hummocky crests. The sands and gravels were fairly well consolidated by the wind so that it would have been fairly easy walking if it hadn't been for those stupid packs. By now my only thoughts, on the sloping surfaces, were: One, two, three … twenty-five. Stop.

At the top of the moraine blanket and just below the level of the entrance to the col itself, we sat for a welcome snack, biscuits and chocolate. ... One of us, playing with the sand on which we were sitting, unearthed a pecten shell, and then another, and a third. I arranged the shells on my ice axe and photographed them, and then said, "Well, I guess some skua must have brought these from McMurdo Sound." (A pair of skuas had adopted us at the base tents.) ...

It wasn't till later, much later, while we were working up our field notes, that we realised the significance of those pecten shells: at some stage a glacier, grounded in McMurdo Sound, was thick enough to plough up the sea-floor bed and carry those shells, and many others, fifty kilometres or so inland. Peter and Dick disagree with this theory and insist the shells are evidence that pectens lived in Wright Fjord in the early Pliocene, three to five million years ago, although I worry about the monstrous sea level change that this implies.

Along the col the sand carried by the persistent winds had eroded the granite boulders into the most bizarre shapes, utterly intriguing to us even in our exhausted state, and the subject of quite remarkable photographs ever since. In other places the ground was covered with bits of dolerite, most about dinner-plate size but some much much larger. These, known as ventifacts, had all been faceted and highly polished into neat geometrical shapes by the persistent wind-driven sandblasting machine. We stopped for a little rest under one particularly cavernous granite specimen.

We spotted another mummified seal, bringing my total to five, and near the highest part of the col – further north it slopes gently down towards McKelvey Valley – we found a very small lake, mostly frozen but with meltwater at the edges. As there was a handy flat spot nearby and the meltwater tasted fresh we decided to pitch tent among the eroded boulders, right there. We weren't in any shape to walk more than five more metres anyway. We just collapsed. Dick had a short

The expedition's campsite on a place later named Bull Pass, between the Wright and McKelvey Valleys. The area was, Bull said, marked by "exotic, sometimes erotic, cavernous boulders" of granite.

air mattress; I made do with my down jacket on top of my rucksack. Pemmican never tasted so good.

The sand attrition continued for most of the night, but the sun shone so that the tent was warm. When the expedition was over the col was named Bull Pass, and hundreds of people have walked through it, worked in it, put up seismic stations in it, and photographed those exotic, sometimes erotic, cavernous boulders. Nevertheless it gives me much pleasure to recall that Dick and I were the first people to camp there! Quite exhausted, we slept well until the alarm rang at 7.45 a.m. so we could get a time check from Scott Base: after all we might have time to take sun shots for an astronomical position determination from our survey station at the top. Hearing Peter's cheerful voice along with the time pips was just about the best thing that happened all day.

I tried to explain to myself – Dick was too far ahead – why my pack was five kilograms heavier than yesterday. My "immediate objective" fell from eighty metres to twenty-five, and soon twenty-centimetre-high boulders required deep concentrated thought lasting several exhausted breaths. Straight up the slope to the west was only 30° or so, and by judicious zigzagging we could keep to the finer scree, where we slipped back less. It wasn't difficult, merely exhausting. Dick was quite a bit faster but was very tolerant and took longer rests, every half hour or less. After five hours we found a moderately sheltered, moderately flat spot for the tent at about 1,600 metres' altitude – the altimeter was one of the victims of the weight purge. There we dropped our packs and continued up the slope, with theodolite and tripod only, as the high cirrus cloud thickened and lowered. After another 150 metres or so of altitude we dumped the theodolite and tripod and continued the short distance to the top unencumbered.

And from the top we gazed down on a minutely set out scene. There was our main base, two pyramid tents and the met screen all neatly laid out, the two ends of the baseline 1,600 metres below – and just about nothing visible above 1,000 metres. We piled invective on whoever was in charge of the weather, made sure the theodolite and tripod were tightly wrapped up for the night, and returned to the tent, mumbling. That really was all very stupid. We'd have been so much better off making a dash – I wondered about that word – up the direct steep slope of the south side of the valley with theodolite, tripod, camera and just a day's food, rather than lug all that clobber round the long, albeit entertainingly scenic, route.

It was snowing as we went to bed before nine o'clock, being unable to think of anything else useful to do. It snowed all night, small and granular, pattering on the tent. When either of us woke we shook the snow off the tent and went back to sleep. We woke for the time pips, not that it mattered. It was still snowing and we weren't surveying. Breakfast at eleven a.m. was half a bar of chocolate, the only food in the tent. After telling me the chocolate was "good and good to eat often", Dick went back to sleep.

I read my book. It snowed. Around four p.m. Dick cooked a meal – lunch I suppose – starting from snow, always a slow process. I tried hard not to be a backseat cook and the meal was fine when it eventually arrived, porridge and scrambled reconstituted dried egg. I cleaned the pots, digging under the snow for some of that remarkably inefficient cleansing powder "Scratcho". Dick wondered if we could market it, using some catchy jingle: "Scratches without cleaning". We played two desultory games of chess on Dick's diminutive board and went back to bed.

Dick told me about Franklin and Coulman Islands, his work with the Forest Service in New Zealand, and of preparing for fieldwork on the west coast by first soaking shirt and socks in a bucket of insect repellent. Barrie and Peter, back on the floor of the valley after being "blown out" of their camp by Mount Odin, asked on the radio whether they should bring us extra food (always their highest priority), which was highly considerate of them but we were fine.

When it wasn't snowing on the following day the visibility was distressingly close to zero at our elevation, although the boys said it wasn't bad in the valley. Dick and I discussed which way, and in what order, the glaciers had moved in this complicated area. I started to read Dick's book *William Shakespeare*, by John Masefield. Although we weren't supposed to discuss the book till we had both read it, Dick said he thought it just a bit too much of a chronicle, without illustrating Shakespeare as a man. I wrote a letter to five-month-old Nicky, complete with a plan of the chaotic tent, jumbled saucepans and sleeping bags. I started to write a deathless piece of poetry about our current state of decrepitude, of which I can now remember only four lines:

And the beauty of that windswept land,
For aeons hid, but now revealed,
That all may wonder and opine,
On nature's puzzles all.

I decided it needed work. Perhaps after dinner. When Peter and Barrie told us that they had found nineteen seals in working the geology thirteen kilometres along the valley, we congratulated them and asked them for the preferred treatment for bedsores.

And the following day made up for it all. It was coldish, –20°C, but no wind and no cloud. Underfoot at the tent there were thirty to forty centimetres of fluffy powdery snow. The glare off the snow was overwhelming and the sun so strong that I had to stop every few metres to wipe the sweat from the inside of my snow goggles, and after 400 metres, thanks to an insulating subcutaneous fatty layer, I was in shirtsleeves, although Dick still had on his anorak.

I carried the light pack and Dick very slowly broke trail, trying to find safe footing on and among the deeply covered boulders. Tricky.

Very. Far away we heard the helicopter droning slowly along the valley, with the gravity meter I hoped, and maybe some mail. I wished I'd left a letter to Gillian to be picked up. Why couldn't we signal the chopper and get a lift to the top of this confounded hill? Three skuas seemed to be having a much easier time of it than we were.

We should have left the tripod standing up but we found it fairly quickly and reached the summit, less than 300 metres above the tent, in two and a half hours, as the unfolding view grew more and more stupendous. It's a huge country. We could see Mount Erebus and beyond, 150 kilometres and more to the south-east; Mount Huggins, in the very impressive Royal Society Range, perhaps 100 kilometres south-east; some huge unnamed mountains to the north; and literally hundreds of minor peaks all around, all decked out in their wedding-dress white. Wonderful. And so exquisitely beautiful we had to pinch ourselves to be sure we were really seeing it all.

A few of the peaks had been given numbers on the aerial photographs, and a subset of these we could identify with certainty. Where do you start with a collection like that? We cleared the knee-deep snow from our intended tripod site, a bit back from the vertical northern face of the summit pile of Peak 16 overlooking Wright Valley, just in case I stepped back to admire our handiwork. Later the peak became Mount Jason, an impressive name to have next to Bull Pass. Dick fished out his camera, a cumbersome but efficient twin-lens Rolleicord, from under his anorak where it was being kept warm, and took a round of black-and-white photographs from the tripod, levelling the camera and then taking a photograph every 30° of azimuth, using his fanciful automatic direction-fixing gizmo. That gave us enough overlap between adjacent exposures to be sure we had photographed everything. ...

I was thankful it was windless up there, but even so my feet grew very cold in the four hours it took to measure the angles to all of the selected peaks. I stopped once to watch Peter and Barrie at Main Base.

They seemed to take a long time over lunch. Down there the flags were showing a wind from the west. Must be a thin wind, because up here it was still calm. I hoped one of the boys had changed the met charts. While he was writing down the angles Dick had made pencil sketches of each of the target peaks. He took another round of photographs, the sun having moved round almost 75° since he took the first set. Last of all he built a cairn, and heartily pleased with our work we ate a couple of biscuits, hard and somewhat like unglazed ceramic tiles. Dick then headed back to the tent to keep the eight p.m. radio contact – he learned from Barrie at main base that indeed the helicopter we had heard had brought the gravity meter and some mail – while I collected a few oriented specimens from close to the top of the dolerite sheet for the proposed paleomagnetic work.

I managed to break the handle of my geology hammer, which ended that particular collecting episode. My spare hammer was down in the tent at main base. Above the dolerite was sandstone and there were bits of it lying around. One of the very light-coloured porous pieces had a dark line just under the edge. Odd bit of weathering, I thought, tossing it away, and thereby missed yet another important discovery, for the colouration was actually a weird endolithic bacterium, as Imre Friedmann from Florida pointed out a decade or more later. Breaking off the rock samples had been nearly as hard as the tombstone biscuits but, to make up, by the time I had reached the tent Dick had coffee ready and also dinner: pemmican and onions, followed by apples, sultanas and more coffee. Great!

When the next morning turned out to be windy, with a low overcast, we realised how lucky we had been the day before, that surveying was out of the question for that day and that we might as well return to main base. The packs were heavier than ever – rocks are denser than food – but gravity was now on our side, and the snow depth lessened as we descended. It was still sufficient to bury any dead seals that might have been around but it did accentuate the terminal moraines along the bottom of the col, laid down, we presumed, by ancient glaciers flowing from the north, and the snow showed up the edges of the huge frost polygons, five- and six-sided and twenty metres or more across.

Under our favourite cavernous granite boulder we made a cup of coffee and ate lunch – cheese and those obdurate biscuits. Eating them, Dick drew blood, cutting his gum, but he was pleased to find that he could use the jagged edge of a biscuit to cut hard frozen butter and cheese.

At the junction of the col and Wright Valley we spent a couple of hours trying to make sense of it all, without much success. Was the moraine deposited by a glacier coming along the col, or up the valley from the coast, or both? There were miles of sand, one stretch in the bottom of the main valley long enough and flat enough to be an aircraft landing strip, except for some frost heaving. Elsewhere were sand and shingle piles that had been wind- and water-sorted, and a few little lakes, domed at the surface because they were frozen solid. Utterly fascinating – and weren't we lucky and privileged to be the first people ever to walk on it?

Following their seven-week exploration of the Dry Valleys the men returned to New Zealand and between them published more than nineteen papers on the geology, biology and geophysics of the region. Colin Bull, who had previously worked in Greenland, later moved to the United States as director of the Ohio State University Institute of Polar Studies. Victoria University of Wellington Antarctic expeditions have continued every year since 1958. In the most recent season, 2014–15, eleven people visited the ice, including geologists, glaciologists, a geophysicist, a geomorphologist – and one science historian.

From *Innocents in the Dry Valleys* by Colin Bull: Victoria University Press, Wellington, 2009. Review of *Innocents in the Dry Valleys* by Terence J. Hughes, quoted in introduction, from *Quaternary Science Reviews* 29: Elsevier, Amsterdam, 2010.

British biologist Bernard Stonehouse at an emperor penguin colony on the Mt Riiser-Larsen Ice Shelf at Amundsen Bay. While most warm-blooded animals stay away from the Antarctic continent during the winter, the emperor penguins remain and breed.

Birds and mammals of Antarctica

Bernard Stonehouse, a British biologist, carried out ecological studies of birds and mammals in Antarctica and researched the impact of human activity on Antarctic plants, animals and soils. In his later years he turned to the impact of tourism. When he wrote this lively piece for a 1965 book about the state of Antarctic science he was working at the University of Canterbury in Christchurch, New Zealand.

The rocks and ice blanket of the southern polar continent offer bare hospitality to animal and plant life. Land vegetation is sparse, restricted by cold, aridity and lack of soil to a meagre assembly of small plants. Algae, mosses and lichens are the characteristic plants of Antarctica; only two genera of flowering plant are known from the fourteen million square kilometres of the continent. Land animals are limited in species to a few invertebrates of soil, vegetation and fresh water; protozoa, mites, tardigrades, nematodes, rotifers, and primitive insects are the only creatures which have so far been found on the continent itself. ...

Antarctic seas are, by contrast, a teeming reservoir of life, comparable in yield per acre to the finest pastureland. At the height of summer their surface waters, although seldom more than one or two degrees above freezing point, contain rich accumulations of microscopic plants and animals, which support, directly or indirectly, huge stocks of seals, whales and seabirds. Debris filtering downward from the surface supports an extensive fauna of sponges, echinoderms, molluscs and other creatures on the ocean floor. Only the intertidal zones of Antarctica are empty of animal and plant life, except in islands north of the region of fast winter ice – for example, South Georgia. The continental shores, with their persistent ice blanket, offer little or no encouragement to the highly specialised forms that elsewhere gain a footing on wave-washed rocks and beaches.

The number of species of animals capable of living in polar conditions is small; abundance in the sea arises from the vast numbers of individuals by which each species is represented, and even on land or in fresh water surprising accumulations of animals of single species may develop in favoured places. Freshwater lakes may teem with rotifers for a few days in each year, small rocks may be black with a dense cover of mites, and hundreds of acres of raised beach may be covered with penguin nests. Land-bound animals tend to be small – the largest animal capable of living entirely on land in Antarctica is considerably smaller than a housefly. At sea a large volume is advantageous to the warm-blooded creatures which dominate the food chains; the world's greatest seals and seabirds are found in Antarctic waters, and the blue whale, the largest animal the world has ever known, grows fat on the summer abundance of food in southern waters.

Marine birds and mammals make only sparing use of the Antarctic continent; with few exceptions they spend more of their time at sea than on land, and feed entirely in the water. Seabirds roost and breed on the shore and coastal peaks. Some may breed far inland; thus snow petrels and McCormick's skuas have been found in small colonies at 2,000 metres in the mountains of Dronning Maud Land, 300 kilometres from the sea. One species, the emperor penguin, breeds mainly on off-lying islands, or even on the sea ice itself. The seals of Antarctica produce their young on shore, or more frequently on sea ice; two species, Ross and crabeater seals, live almost exclusively on floating pack ice and are seldom seen on or near land.

Whales are entirely aquatic, but some species show a tendency to gather in sheltered inshore waters where, often enough, plankton is locally enriched. Like many of the seals and seabirds, whales migrate southward in spring and summer, when the surface waters of Antarctica provide their greatest abundance of food. They winter in lower, warmer latitudes, where food is usually scarce but heat losses from the body are smaller. Unlike seals and birds, whales produce their young in the warmer part of their range, and the summer excursions to high latitudes are periods of feeding and fattening rather than of breeding.

Forty-three species of birds and six species of seals breed in the area generally defined as the Antarctic region – that is, on islands and coasts within the bounds of the Antarctic Convergence. Only twelve species of birds and four species of seals breed on or near the shores of the Antarctic continent. Of the factors that combine to restrict the number of species in the far south, isolation has probably played an important role throughout the geological history of the continent. At present the narrowest gap separating Antarctica from the rest of the world is the Drake Passage, a 650-kilometre stretch of deep and tempestuous water that forms a likely barrier to colonisation by modern plants and animals. Land bridges may have spanned the gap (possibly through the islands of the Scotia Arc) at intervals during early and middle Tertiary times, when the climate of Antarctica was less rigorous than at present. The continent supported temperate or subtropical plant life and insects in the past, but no fossil remains of amphibia, reptiles or land birds and mammals have been found. The only vertebrate fossils so far recorded are of Tertiary penguins and whale-like mammals, sea creatures akin to those that inhabit Antarctica at present; it is possible the vertebrate fauna of the continent has always been restricted to animals that could reach it by flying or swimming. The seabirds and seals of today owe much of their success to Antarctica's isolation, for they flourish in the complete absence of large mammalian land predators.

Penguins and seals probably spend more than half their lives in cold water, an exacting environment for which they are well equipped. Their ability to live on the coldest continent follows almost incidentally, for adaptations to the one environment confer benefit in the other. Marine animals generally, and the warm-blooded vertebrates in particular, are more abundant in cold than in warm water, for cool seas, including the cold-water masses and currents that penetrate toward the tropics from polar and temperate latitudes, are richer in food. To take advantage of food resources in cold water, warm-blooded animals need the means of reducing heat losses between their bodies and the sea. The dense, water-repelling plumage of penguin and petrel, the thick tough skin of whale and seal, the fur borne by some of the seals of temperate and low Antarctic latitudes, the subcutaneous fat or blubber common to all warm-blooded marine vertebrates, are devices for reducing the flow of heat between warm animal and cold sea.

Compactness and size are further adaptations in the same cause. The extremities of seals and penguins are generally short and bony; muscles are concentrated in the body mass, so little blood need circulate in the periphery. Largeness is itself an asset, for large animals have a low ratio of surface to volume and can therefore conserve heat more efficiently than smaller animals. Some animals, notably the whales, have a heat exchange system in the blood vessels of the skin, whereby cooled blood flowing inward from the skin is warmed by arterial blood flowing outward. Seals and sea birds capable of living in freezing water are apparently undismayed by air at very much lower temperatures, possibly because of the lower conductivity of still air. A cold windy spell of weather in Antarctica sends seals rapidly back into the sea and may delay by a few days the onset of breeding in some birds; to this extent the animals are clearly responsive to cold. They are capable, however, of tending their young through blizzards lasting several days, and rarely seem to die of cold when food is plentiful. Young birds are more vulnerable than adults, and a single blizzard may give rise to heavy losses in a colony containing nestlings.

The warm-blooded animals of Antarctica seldom experience extreme cold. Most of them keep away from the continent during the coldest months; only the emperor penguin remains to breed in the depths of winter. In thinking of Antarctic conditions man takes the biased view of a mammal accustomed to living in temperate regions. Poorly adapted for polar life, he experiences hardships and difficulties which few Antarctic animals have to share.

Visiting continental Antarctica only in summer, and living near the coast where mean temperatures are ameliorated by the warmth of the sea, most southern seabirds and seals avoid the extreme conditions that man encounters on barrier and inland ice. Few birds fly inland; in their summer wanderings they are unlikely to encounter temperatures as low, for instance, as those met every winter by birds of eastern and central North America. Their own winters are spent among the pack ice or in warmer regions beyond; only when mean monthly temperatures are approaching or rising above freezing point will they return to breed in the region of the continental coast. Their environmental temperatures in summer are in any case higher than climatic records show, for the warming effects of sun on rock, and sheltered conditions near the ground, affect considerably the microclimate in which they live during their months in Antarctica. Seals also miss many of the conditions men encounter, for they lie in sheltered positions on the sea ice and return to the comparatively warm sea when temperatures are low or winds strong.

The continuing success of Antarctic mammals and birds, their numbers and wide distribution about the continent, above all their many remarkable adaptations for their environment, make them subjects of particular interest to biologists. The recent spate of scientific activity in Antarctica has provided many opportunities for their study, and the present account is an attempt to summarise some of the more important findings of research in the last quarter-century.

From "Chapter Six: Birds and Mammals" by B. Stonehouse, in *Antarctica* edited by Trevor Hatherton: A.H. and A.W. Reed, Auckland, 1965.

A meteorite found by ANSMET – the US Antarctic Search for Meteorites
– at Szabo Bluff in Antarctica's Queen Maud Mountains, 2012.

Catching falling stars

The first meteorite ever found in Antarctica was a one-kilogram chondrite – a non-metallic stony meteorite – discovered in Adélie Land by Douglas Mawson's 1911–14 Australasian Antarctic Expedition. The second and third were discovered in 1961, one in Adélie Land and the other in the Thiel Mountains, and the next was found in 1964 in the Pensacola Mountains of Queen Elizabeth Land. American meteoriticist William A. Cassidy summed up the situation: "At first glance there was nothing to recommend the Antarctic continent as a place where one could find great numbers of meteorites, since so few had been found." This changed one day in 1969, when a team of Japanese glaciologists came upon nine more meteorites at the Yamato Mountains in the Queen Maud Land region.

The Yamato mountain range wraps the ice sheet around its shoulders like an old man with a shawl. Ice coming from high off the ice plateau of East Antarctica, arriving from as far away as a sub-ice ridge 600 kilometres to the south, finds this mountain range is the first barrier to its flow. The ice has piled its substance up against the mountains in a titanic contest that pits billions of tons of advancing ice against immovable rock, whose roots extend at least to a depth of thirty kilometres. The ice is moving because billions of tons of ice are behind it, pushing it off the continent and into the sea. Ultimately it yields, diverging to flow around the mountains. On the upstream side the rocks have been almost completely overwhelmed – only pink granite peaks protrude above the ice, which spills down between and around them in tremendous frozen streams and eddies, lobes and deeply crevassed icefalls.

The change in elevation of some 1,100 metres between the high plateau upstream of the mountains and the lower ice flowing away from the downstream slopes creates a spectacular view of this giant downward step in the ice surface. Almost constant howling winds from the interior blow streamers of ice crystals off the mountain peaks and "snow snakes" dance down the slopes in sinuous trains, as if somehow connected to each other. The scale of the scene is such that people become mere specks in an awesome, frigid emptiness.

In 1969 a group of Japanese glaciologists were specks in this scene. With all their supplies and equipment, they had travelled inland 400 kilometres from Syowa Base on the coast, to reach the Yamato Mountains (called the Queen Fabiola Mountains on most maps) and carry out measurements on the velocity of ice flow, rate of ablation and ice crystallography. Their safety depended on the reliable operation of two tracked vehicles in which they ate, slept and waited out the storms. These scientists were physically hardy and highly motivated. Because the Japanese supply ship could reach Syowa Base only in the middle of summer when parties had already left for the field, they had already wintered over at Syowa Base and would spend another winter there before being able to return to their families, just so they could spend the four months of Antarctic summer at this desolate place, gathering fundamental data along the margin of a continental ice sheet. One of them, Renji Naruse, picked up a lone rock that was lying on the vast bare ice surface and recognised it as a meteorite.

In the preceding 200 years only about 2,000 different meteorites had been recovered over the entire land surface of the Earth, and finding a meteorite by chance must be counted as extremely improbable. It's lucky, therefore, that this initial discovery at the Yamato Mountains was made by a glaciologist, who would not be expected to have a quantitative understanding of exactly how rare meteorites really are, and of what a lucky find this should have been – Naruse and his companions proceeded to search for more. By day's end they had found eight more specimens in an area of ice five by ten kilometres – a tiny, tiny fraction of the Earth's land surface.

Until that time, such a concentration always represented a meteorite that had broken apart while falling through the Earth's atmosphere, scattering its fragments over a small area called a strewnfield. In such a case all the fragments are identifiable as being of the same type. In this instance, however, all nine meteorites were identifiably different and so were from different falls. A meteoriticist would strike his forehead with the palm of his hand in disbelief.

Naruse and his companions undoubtedly were pleased with this unexpected addition to their field studies but there is no record that they immediately attached great significance to the find. They bundled up the specimens carefully for return to Japan, and then resumed the ice studies that had drawn them to this spot. The ice at the Yamato Mountains, however, was destined for great fame not for its glaciology but for the thousands of meteorites that would later be found on its surface. One might say that the Yamato Mountains icefields were *infested* with meteorites.

Inspired by the Japanese discovery, Cassidy launched a meteorite-hunting expedition to the Transantarctic Mountains in 1976 – the first of fifteen seasons he would spend on the ice – and co-founded the US Antarctic Search for Meteorites (ANSMET). Through the efforts of ANSMET and similar programmes run by Japan and China, there are now more than 30,000 Antarctic meteorites stored in research bodies such as the Smithsonian Institution in the United States and available for scientists to study. These meteorites include lunar and Martian meteorites, rare stony-iron meteorites, and the more common chondrite rocks that come from the asteroid belt.

William A. Cassidy

It was Cassidy who first realised that – given the quantity and range of different types of meteorites found in Antarctica – there must be a concentration mechanism at work. Here he explains this process and why Antarctica is such a great place to find meteorites.

There are two reasons why concentrations of meteorites can occur in the cold desert that is Antarctica: the cold temperatures and lack of exposure to liquid water greatly inhibit the process of chemical weathering, so meteorites can be preserved for great lengths of time; and the slow movement of the ice sheet toward the coastline sometimes ends prematurely at a surface where ice wells up against a barrier and disappears by ablation. In such an area the arriving ice disgorges any meteorites it has collected during its travels and leaves them stranded on the ablation surface. Because they do not weather very fast, even when exposed to the cold dry air, meteorites can become concentrated on ablation surfaces if the local conditions leading to ablation have persisted long enough. So in Antarctica the process of meteorite concentration can occur both across a distance, represented by the upstream path length of the ice, and also through time, made possible by the slow weathering rates. ...

Suppose the meteorite falls near the centre of the continent at, say, 3,500-metre elevation. It initially buries itself in the snow. During succeeding seasons more snow falls, burying it more deeply. When it is about thirty metres deep the snow begins to undergo pressure recrystallisation and to form ice, squeezing out a lot of the air between the grains. The meteorite is now embedded in ice.

As time passes and more snow falls, the meteorite follows an ever deeper trajectory and moves outward with the ice toward the edge of the continent. In this way, most meteorites that fall on to the ice sheet eventually are buried at sea. What a terrible waste of scientifically interesting material! The exceptions, which are the ones we find, are

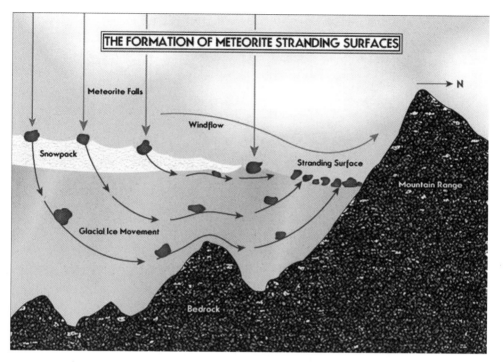

When meteorites fall, they become embedded in packed snow, which eventually becomes glacial ice. Over time, the glaciers flow towards the sea. As they try to push up and over mountains in their path, they act like giant conveyor belts, carrying the meteorites to the surface. Over time, wind exposes the meteorite concentrations at places called "stranding surfaces", as this diagram by Chicago's Field Museum shows.

associated with patches of ice that cannot reach the sea because they are trapped behind a barrier.

Ice has nowhere to go if its path has led it to a barrier, such as a mountain range, that it cannot flow over or around. If there are sufficiently high winds that blow pretty constantly at this site, the overlying snow layer is stripped away and the ice is exposed. Now two things happen: the constant wind blows ice crystals over the surface, producing a sandblasting effect that wears the ice away; and in the summer, solar heating causes vaporisation at the ice surface, which also wears it away. The wearing-away process is called ablation. So even though ice trapped behind a barrier cannot flow away, it can leave by ablation.

It is important now to understand that ice acts like a sluggish hydraulic system: as surface ice leaves by ablation, upstream ice moves from the interior of the continent into the stagnant zone to restore the original surface elevation. So essentially, in a zone of stagnant flow, ice can leave by passing through the ice–air interface, and in the absence of serious climate change that would upset this equilibrium the supply of new ice arriving at this interface always is just the amount that has left by ablation. The significance of this to meteoritics is that if an absolute barrier to ice flow exists for a very long period of time, it is certain that the very occasional meteorite will be brought up by the ice and will be left behind, stranded on the ice surface. For this reason, we refer to these patches of ice as meteorite stranding surfaces.

Now, because meteorites weather away so very slowly in Antarctica, it is easy to see how a concentration of meteorites can be created on a meteorite stranding surface. The essential ingredients are stagnant ice flow, high winds and very large chunks of time. Because the stranding surface must have been there for a great length of time, it may have accumulated meteorites that fell directly on to it, in addition to those carried there by the ice. ...

Over the years we have found that complete stagnation of ice flow is not absolutely necessary to form meteorite concentrations. Most of the occurrences we have discovered, in fact, are not located behind perfect barriers to flow. In many field situations the ice of the stranding surface seems to be descending a step and flowing away. We believe this downward step is caused by ice actually overwhelming a subsurface barrier that is not high enough to completely restrict flow. In these cases, ice is flowing over the barrier and still a concentration of meteorites is found on the surface. It seems necessary to assume that the barrier slows the ice down and high winds cause ablation fast enough to allow a concentration of meteorites to accumulate at these places.

From *Meteorites, Ice, and Antarctica: A Personal Account* by William A. Cassidy: Cambridge University Press, Cambridge, 2003.

Lloyd Spencer Davis with an Adélie penguin. Beginning here at Cape
Bird on Ross Island, the zoologist has spent over thirty years studying the
behaviour of Antarctica's most populous penguin.

The accidental penguin biologist

Lloyd Spencer Davis first travelled to Antarctica in 1977. His original PhD proposal to study Weddell seals in McMurdo Sound was not accepted, and at short notice he switched to a project on penguins. A behavioural biologist, Davis gathered data by painstaking, long-term observation of his subject animals. Studying penguins became the focus of his scientific career, and in the course of three decades of research on Antarctica's Adélie penguins he published one of the most comprehensive sets of observations of any bird's feeding, courting, mating and nesting habits. As this story opens Davis and his field assistant Max Wenden have just been dropped at the northern extremity of Ross Island, Cape Bird, home to 60,000 Adélie penguins, a green hut and a radio tower.

I sat mesmerised. My ambition to go to Antarctica had always been about adventure in what I imagined was a harsh and colourless landscape. I had never dreamed that Antarctica's most striking features would be its beauty and its colour.

A few hours earlier the helicopter had dropped us next to the hut, saving us the exertion required to carry nearly a ton of gear (stuff required for survival, stuff required for eating and stuff required for the science I intended to carry out) from the usual helo pad, located on the beach some thirty metres down a steep incline from the hut and half a kilometre away.

Once the penguins began breeding, the latter would be our only option. Actually "hut" is far too glorified a term for what was to be my abode for the next three and a half months. As a child, I'd built tree huts that were more substantial. It was painted in the ubiquitous Scott Base green – as if the New Zealand Antarctic Division had got a discount at Dulux as long as they took what was likely to be the least favourite colour in their inventory: a sort of dark forest green that was far too bold for any house – and, to be honest, any hut.

The hut, such as it was, consisted of an entranceway, which also served as a refrigerator by virtue of its unheated status; a single living room that contained six bunks, the tiniest of cooking spaces, a table, three chairs and two windows; and an adjoining small room that served as a laboratory for the scientific work and as a semi-private place to wash ourselves, and also contained a stove that was used to melt ice and snow for water and to keep the hut warm. The entire structure was made with two-centimetre-thick plywood. No insulation at all. Nada. The icicles crept up the walls beneath the bunks even with the heater on. The term "windows" also suggested something more than they were: they were simply holes cut in the plywood over which a piece of clear Perspex had been screwed. It didn't qualify as single glazing, let alone double glazing.

Yet I fell instantly in love with the hut in a way I have not with any place I have resided in before or since. It had a rustic charm that appealed to that part of me that sought adventure in a harsh and hostile environment called Antarctica. The hut had been built by researchers from the University of Canterbury in the mid 1960s. Originally intended as only a temporary shelter, the fact it was more substantial than a tent meant that it was retained in the absence of alternatives. One of its first occupants had been a chap called John Darby, judging by the carving in the bottom of the bunk above mine: "John Darby 1965–66, 1966–67, 1967–68."

"Slow learner," Max had said, in what transpired to be his one attempt at humour in the two weeks he would accompany me.

I left Max trying to get the heater going – the prospect of being without heat in such a flimsy hut in such a non-flimsy environment was not much of a joke either – while I headed out to survey the colony for the first time. It was just after eleven p.m. and on the far side of McMurdo Sound the sun sat poised atop the Royal Society range of mountains that formed the horizon for us to the north-west. The clouds in the western sky were already a brilliant orange that contrasted spectacularly with the deep blue in those patches devoid of clouds.

On leaving the hut I walked 100 or so metres to the east, over a small headland known as Robbers' Roost because it was a favourite hangout for skuas as they surveyed the penguins for any opportunity to steal an egg or a chick. It was also an ideal vantage point for a hitherto rat biologist and wannabe seal biologist to survey the study area related to his latest incarnation. Penguin biologist: it didn't seem quite real yet. I hadn't even seen a penguin.

And then I did.

I could make out, even without my binoculars, the unmistakable black-and-white figures of waddling penguins on the gently undulating slopes of the Northern Colony at Cape Bird. The colony occupied the floor of a massive amphitheatre that was backed by the Mount Bird ice cap, including a large glacier tongue that careened down the slope and into the sea – if careening can be said to be a characteristic of any glacier. This one looked so formidable and rugged, serrated by huge crevasses and decorated by monstrous icicles that hung from its sixty-metre-high lip, I did not imagine for a moment that it hung about, moving sedately. It looked like it took no prisoners. It looked like it careened.

The penguins seemed tiny from the distance of Robbers' Roost. In all, I counted fourteen. Perhaps there were others I could not see from where I stood, but surely I had arrived at the colony ahead of the vast majority of the many thousands of penguins, and importantly before the process of breeding had begun.

The sea itself was a mosaic of ice floes. I had read that in some years the sea ice that forms during winter did not break up until much later in the season. But there had clearly been a storm or other event that had smitten the sea ice into a giant jigsaw consisting of thousands, millions, perhaps billions of pieces. Truly I did not know how many: there were lots, as far as the eye could see. When flying down in the Starlifter I had managed a glimpse through one of its few windows, and the jigsaw puzzle nature of the pack ice was unmistakable from 10,000 metres. The individual floes before me now, these pieces of pack ice, banged together and jostled, drifting surprisingly quickly with the current and the wind, although there was precious little of the latter. And on some of the ice floes, there were clusters of penguins.

Clasping my camera and binoculars, I half ran, half slid down the volcanic rocks of the headland in my yellow mukluks as I headed directly to the beach, my yellow down jacket unzipped because at minus sixteen degrees Celsius and with little wind, the temperature felt positively benign.

The interface between land and sea was unlike anything I'd ever encountered before. Winter's storms had piled blocks of ice up at the sea's margins, creating a wall of ice blocks referred to, appropriately enough, as "push ice" because the storms and the sea had pushed it there. I clambered over the push ice, getting as close to the water as I dared. The push ice formed something akin to the walls of a swimming pool so this was not like your average beach: the water was instantly deep. It being very clear, I could see the stony bottom and figured it had to be about six metres deep. This was more a beach suited to diving or drowning than to paddling.

I sat mesmerised.

I listened to the gently slapping water as the ice floes rose and fell ever so slightly, like the sea was breathing – the sighs of a sleeping sea. The ice floes had been transformed into a fleet of golden-painted boats by the reflections from clouds that were now a fiery red as the sun slid out of sight behind the Royal Society Range.

A small ice floe with perhaps a dozen penguin passengers drifted to within twenty metres of me. The penguin closest to the edge of the floe had been sleeping but now raised its head. A moment later there was a loud "bluuub" and a dark, hideously huge head reared out of the water at the penguin. It looked like T. rex. It looked like the Loch Ness monster on steroids. The leopard seal slid back into the water only to reappear at the far end of the floe, rearing perhaps a metre and a half out of the water. Mind you, at least half a metre of that was just head. I watched the leopard seal do the same at other floes before I lost sight of it.

I sat mesmerised.

I sat there, reflecting that this had to qualify as the most beautiful moment of my life. Among the interplay of sun, ice, penguin and seal, I had been privileged to glimpse something fundamental about the world – about the good bits of life, about the big questions that sometimes eluded me.

"Bluuub." With a hiss of expelled air and the crunching of the small pieces of ice that carpeted the sea's surface, the leopard seal's huge head rose up directly in front of me, its nose an arm's length from my own, its nostrils flared, its stare intense.

I sat mesmerised, me playing snake to its mongoose. Me with heart in mouth, but a heart that beat faster not just with fear but with awe too. I was close enough to exchange breaths with a perfectly designed killing machine and I couldn't help but notice the beauty in that.

From *Professor Penguin: Discovery and Adventure with Penguins* by Lloyd Spencer Davis: Random House, Auckland, 2014.

Food chain

The big chicks chase any adult
who might be a parent.

A bird waddles back from the water,
belly crammed with krill,
then drops and toboggans,
tired after all that travel,

and pauses just
at the edge of the nursery

*

where at once it's surrounded

*

while the skuas go on cruising the colony:
dangerous wings
slicing in from the edge

*

and out at sea (shouts someone)
a leopard seal has taken a penguin …

*

For a while, we watch the seal
play with the life it will eat;
for a while, it seizes with teeth
then tosses the bird somewhere behind;
then turns and seizes again
and tosses it to and fro, to and fro …
to and fro for a bit …

*

and for a while Mike goes into close-up
at the edge of the cliff, filming it.

Bill Manhire

Tiny krill, most of which reach only one to two centimetres in length, are the key food for the world's largest living mammal, the blue whale, and swarming in their millions comprise one of the largest biomasses of any animal on Earth.

Krill

In *The Crystal Desert* American ecologist David Campbell wrote about his experiences living and working at a Brazilian research station in Admiralty Bay on King George Island, near the tip of the Antarctic Peninsula. Campbell described Admiralty Bay, a three-fingered fiord on the south-east side of the island, as a microcosm of Antarctica – "it is oceanic, it is terrestrial, and its heart is glaciated, but during the summer its shores are warm and rainy." During the summer the bay is home to humpback whales, elephant, leopard and Weddell seals, Adélie, gentoo and chinstrap penguins, giant petrels, skuas and terns, and a bustling population of international researchers and civilians.

Campbell's main scientific interest was the tiny creatures living in the icy water. On this starry summer night in 1987, he puts on his waders and enters the sea. Laden with flashlight, net and collecting bucket, he walks out into the shallow bay to where a floodlight on top of a post is being used as a beacon to attract plankton. In the water beneath the light, he finds "a blizzard of plankton", most less than a centimetre long. Using his net, he deposits a sample of this "living bouillabaisse" into his bucket, returns to shore and enters his refrigerated laboratory.

There is only one short link in the food chain between a diatom and a one-hundred-ton blue whale – between one cell and the largest of all animals – and that link is the Antarctic krill. This may be one of the shortest and simplest food chains on Earth, but the numbers are astounding. An adult blue whale eats three to four tons of krill per day over the four-month Antarctic summer: in all, about a half billion individual krill. Each krill that perishes in the maw of the whale was

David G. Campbell

wrought according to a genetic blueprint that is unique in the universe. And these half billion krill, in turn, have during their lives filtered from the sea perhaps ten billion billion diatoms, each a mote of the sun's energy and also an evocation of self. This is far more diatoms than there are visible stars in the sky tonight. As far as we know, that number is about the same order of magnitude as the number of stars in all of the cold, invisible corners and limitless planes of the universe.

I net a single krill, three centimeters long, from the vortex in the bucket, place it in a petri dish of chilled sea water, and examine it under the low-power dissecting microscope. A few preliminary checks tell me what species it is. I note that the eye is spherical, not lobed. This tells me that it is not in the genus *Thysanoessa*, the big-eyed krills, but must be in the genus *Euphausia*. The animal has no dorsal spines on its abdominal segments, eliminating the spiny krill (*E. triacantha*) and the northern krill (*E. vallentini*). This is no surprise, since both of these species are largely subantarctic, abundant north of the Antarctic Convergence and in the coastal waters of Patagonia and the Falkland Islands, but rare in Admiralty Bay.

The krill under the microscope has a stubby blunt spine protruding from the thorax between the eyes. This detail informs me that I am peering at either an *E. superba* or a pygmy krill, *E. frigida*, and not the ice krill, *E. crystallorophias*, which is also abundant under the light tonight. The options are now only two. I examine the antennae. This krill has a bilobed lappet on the first segments of its antennae, as opposed to a triangular lappet. The whole process of keying out this animal has taken only a few seconds. It is an *E. superba*.

Most people lead their lives indifferent to *Euphausia superba*, the Antarctic krill. They don't know what they are missing – and not because someday soon their pizza may be garnished by krill. There is a greater biomass of *E. superba* – about 600 million tons – than of any other species of animal on Earth; more than humans, more than

240

the fleets of albacore tuna that ply the Pacific, more than the locusts that periodically devour equatorial Africa. There are enough krill to sustain the indifferent predation of the baleen whales, the burgeoning population of crabeater seals, the porpoising penguins, and the darting antifreeze fish. The DNA of krill has gone berserk, stamping out countless copies of itself. I am now examining a single individual of this most biomassive of all animals, one copy of unknown trillions, each sufficiently identical for me to key it to species but each, of course, unique, with its own set of recombinant genes, its own suite of serum proteins, its one particular morphometric design.

My E. *superba* is like an animate crystal refracting the hot microscope light; its clear shell scatters the beam into a partial spectrum of blues and a hint of orange. The segments – head, thorax, seven tail sections – interlock like transparent armour plating and give the body a clanking rigidity. The shell is chitinous, made of a carbohydrate related to cellulose. All of life's functions must take place within this vessel; to grow, the krill has to split its shell asunder and cast it away, replacing it with a new shell lying nascent and unhardened beneath the old.

The thorax, which makes up a third of the krill's body length, is the largest segment. Forward, nestled in thoracic sockets, are the stalked compound eyes, brown with light-absorbing pigments. They are faceted beyond the resolution of my eye and brain. The krill pulses its thoracic legs, three pairs of short endopods that push water past its gills and three pairs of extended, bristled exopods that cup into the thoracic basket, a filter that sweeps the water of diatoms and other plankton. The filtering legs of the thoracic basket are to the krill what baleen is to a whale. Their bristles look like fibers of spun glass. Behind are six pairs of short pleopods, each pair originating from one of the abdominal segments. The seventh pair of legs emerges from the final segment and expands into the tail, spreading and contracting like the fan of a flirting geisha.

David G. Campbell

The thelyctrum, the seminal receptacle at the base of the thorax, tells me that this is a female. But she is virgin. The thelyctrum bears no spermatophore, the chitin-wrapped packet of sperm that would have been placed there during a lover's brief embrace as the couple drifted among the hordes of others in the open sea. Her ovary, a smudge in the translucent thorax, is tiny and barren. This makes sense, since spawning now in Admiralty Bay, with its many predator-infested shallows, would be premature. Better to wait, to drift to the edge of Antarctica, the edge of the ecosystem, to release the depth-seeking eggs in the southward-drifting subsea currents. She has been feeding: her stomach and long gut are coloured green-brown by diatoms. And her translucent heart is pulsing so fast that it appears to be fibrillating. The blood is clear and invisible.

The krill is gyring wildly in the captive ocean but getting nowhere. Then, in desperation, she snaps her tail and lurches violently backward until at last she becomes stuck on a strand of bright green algae. The sensory flagella at the tip of her antennae quiver slightly, tasting the water with alarm: oxygen levels are dropping, the water is warming up.

Holding her in the forceps, I insert a thin hypodermic needle between the segments of her tail and extract a drop of blood, which I place on a slide under the compound microscope. First, 10x. Like the sea water that swarmed with diatoms, the blood eddies in its own miniature cosmos. The krill's blood cells, known as hemocytes, are spherical globs only three or four microns long that sweep past in uncountable numbers. But as I watch, constellations of cells distill from the torrent and freeze on to the slide. Within seconds, more cells cling to the slide and the serum becomes viscous. All movement stops, like the still frame of a motion picture.

100x. I can see the individual cells now, stuck to the slide. Some are granular, others are clear, having leaked their cytoplasm on to the slide in lobed blebs. Some of the hemocytes explode before my eyes, leaving

242

A krill's eye gives a clue to its species. The spherical eyes of these ones indicate they are of the genus *Euphausia*.

a cell membrane like a punctured balloon. The krill's hemocytes are her first line of defense against pathogens. They intercept invading viruses, bacteria and protozoans, rupturing and leaking their cell sap on to the invaders, clumping on to them and causing them to disintegrate. The hemocytes recognise other living things as foreign. This is not an easy matter on the scale of cells. They also recognise the non-living slide, which is covered with a scum of life – the metabolic by-products of bacteria deposited from the air, grease from my fingers, cells from my breath – as foreign. By recognising what is other, the hemocytes must also recognise self, the krill's personal genetic code. They declare each being's uniqueness in the wide sea.

I turn my attention back to the krill in the petri dish. During the minutes that I was peering at her blood, her legs have ceased pulsing, her heart has stilled. The krill is dying in the heat (a few degrees is all it takes) and her clear tissues have turned opaque, like a spreading cataract. Her proteins, adapted to subfreezing cold, are coagulating, folding, twisting, and binding on to themselves, assuming shapes that are useless to life's processes. The crystalline animal transforms into an opaque gray corpse, like a shrimp steaming in a boil. I begin tearing her apart with dissecting needles, probing her tissues for hidden parasites. This is my science, my reason for being here on this cold starlit night, seeking the information that this animal can only posthumously yield.

The krill in my petri dish, this trivial but unique piece of protoplasm, has led a complex and wonderful existence. She is probably no more than a year old, and with extraordinary good fortune she could have lived up to seven years, which is long by the standards of zooplankton. In her short life she has travelled through hundreds of kilometers of open sea and has traversed the vertical layers of the Southern Ocean. Krill are heavier than water and sink if they do not constantly beat their six pairs of swimming legs, a necessity that makes every one of life's activities – hunting, eating, even resting – expensive. This krill never touched the sea floor, never took refuge from the current beneath a rock nor clung to a strand of kelp. Other Antarctic zooplankton, such as copepods and amphipods, manufacture buoyant oils and waxes that keep them from sinking, but krill must constantly swim.

These wonderful peregrinations are all controlled by instinct. The fleeting electrical software that directs these behaviours is locked inside a few bundled neurons behind her stalked eyes. Yet this smudge of cells knew the waxing and waning of the pack ice and the drift of water masses the size of continents.

The Southern Ocean, pushed ever eastward and nudged into gyres and eddies by the West Wind Drift, is an unpredictable place to be a plankter. But this krill took advantage of the general patterns of ocean movement. She was probably hatched near the Antarctic Convergence, hundreds of kilometres out to sea, although she could have come from one of the smaller populations that breed close to the continent and even in Bransfield Strait. In any case, her conception took place over deep water, and the egg that was to grow into this krill immediately sank to the lightless depths, an adaptation to avoid predators in the life-filled top of the sea. A female krill may spawn as many as sixteen times during the course of the austral summer, and each batch of eggs, depending on the size and age of the female, may number as many as 2,000. Such extravagant egg production is typical of planktonic animals, which try to overwhelm a hungry sea with sheer profligacy.

As the egg descended into the depths of the sea, its cells divided until it had the appearance of a microscopic raspberry. By the time it reached its nadir, 2,000 metres below the surface, it was an embryo with limb buds. In this dark cold water, at a pressure of nearly 200 atmospheres, the egg hatched into a six-legged mouthless larva known as a nauplius. Still relying on the energy from its yolk sac, the nauplius reversed its migration and slowly journeyed back toward the sunlit surface. And as it rose it repeatedly shed its confining shell, grew and metamorphosed. The nauplius moulted three times before transforming into a spiderlike calyptopsis larva. The calyptopsis moulted three more times before transforming into yet another life stage, the furcilia, a shrimp-like juvenile that for the first time resembled an adult. That transformation took place just as the krill arrived at the surface. The furcilia is the first life stage endowed with a mouth and full motility, and this krill began to feed voraciously on its fellow wanderers in the planktonic fields. The slow descent and return took from two to four weeks. Besides allowing the krill to evade predators, the deep passage served another purpose. By sinking, the egg entered the Antarctic deep water, which carried it south to the edge of the continent, and the furcilia emerged in the nutrient-rich upwelling of the Antarctic Divergence just when it was time to eat. By migrating vertically through the shearing water masses, the krill instinctively knew how to hitch rides in the currents. By rising, it travelled north; by sinking, it travelled south.

If I hadn't duped the krill with the light and caught it, the chances are overwhelming that something else would have. Only a small fraction of one percent of the eggs spawned each season survive the journey to adulthood or live to the unlikely old age of seven years. The survivors do not remain scattered over the wide sea but are gregarious and congregate in swarms. Krill swarms are among the greatest biological phenomena on Earth. They are the planktonic equivalent of a redwood forest, of migrating monarch butterflies, or of the migration of wildebeest and zebras across the Serengeti. On March 7, 1981, a "super swarm" of krill was detected near Elephant Island by the side-looking sonar aboard the RV *Melville*. The swarm had a surface area of approximately 150 square kilometres and was at least 200 metres deep. Some samples netted from the swarm contained up to 10,000 Antarctic krill (equivalent to ten kilograms) per cubic metre; yet other samples, taken minutes later and in the same spot, contained almost none. The scientists aboard the *Melville* estimated that the swarm contained between two and nine billion kilograms of Antarctic krill. It also contained other zooplankton: lesser amounts of two other species of euphausids and five species of copepods, as well as amphipods and salps.

Why do krill aggregate in such astronomical numbers? The question is all the more perplexing because it is almost certain that the most important predators of krill, the baleen whales, could not feed efficiently if krill didn't swarm. (Nor, for that matter, could Russian trawlers economically catch krill.) If krill were solitary, then whales would expend more energy seeking and catching them than they would ever derive from the krill as food.

The reasons for the aggregations become clear when one swims through a swarm of krill. The swarm is not homogeneous; instead, the krill congregate in layers, ribbons or undulating curtains, like an aquatic aurora australis. Nor are they randomly drifting in these aggregations. Most of them are oriented in the same direction relative to one another, each probably taking cues from the subtle wakes and eddies of the others. Since krill must swim to keep from sinking, the swarm progresses in concert through the water. As they swarm, they feed, leaving behind a plankton shadow, a lifeless zone of filtered water. And this may be the most important clue of all. Like the krill themselves, the blooms of phytoplankton on which they feed are patchy in distribution, and a krill in a swarm is more likely to detect pockets of plankton than if it were solitary. The swarm, in effect, increases each krill's sensory range. As one krill darts toward the plankton, and then another and another, the signal may ripple through the swarm like a billow of wind over at field of grain. Finally, the swarming of krill, like the schooling of fish, confers a certain anonymity on each individual, which confuses predators that must hunt krill one at a time, like squids and penguins, as opposed to the mass-feeding baleen whales. Swarming provides a trade-off: enhanced feeding on plankton and protection from certain predators, but increased vulnerability to others.

From *The Crystal Desert: Summers in Antarctica* by David G. Campbell: Minerva, London, 1993.

Small fry*

A great mouth opens: we flip backwards

out of our skeletons. Better to   abandon ship. Safer to

leave them drifting: our stellar cores

a lace of ghosts   ice picks.   We are meal rowing away

from its cutlery   we are ice hightailing

from its cloud   —   we are grooming   legs   feeding

legs   swimming legs, five sets   it's true, friends: we are

mostly legs   therefore direction   therefore freedom   and our freedom

can be seen   pulsing   translucent in our hearts   and in the waving

of our bristles.   We swim slowly but   we swim ceaselessly   to keep

from falling   to keep from disappearing   —   we have never   drifted.

We are not the sea's   plumage   but its innards   but the stars

it wears.   We feather   our oars in it   we fill it with eyes   held up

on stalks   we haul the sea into us   like a net hauls   fish   and at the end

of summer   when we grow   back into children   it is still

at our mercy.   We fill it with our mouths   and tails   the hard parts

of which   reflect sound

Ashleigh Young

* This poem was inspired by the statement "The prevailing view of krill as zooplankton leads to the assumption that they merely drift in the currents and that there is a 'conveyor belt' carrying krill around the continent" by Elizabeth Leane and Steve Nicol in "Charismatic Krill? Size and conservation in the ocean" in *Anthrozoös: A multidisciplinary journal of the interactions of people and animals*, volume 24, issue 2, June 2011.

Palaeontologist John Long searching for fish fossils on Gorgon's Head, a peak in East Antarctica's Cook Mountains.

An average day in the deep field

"No one goes to Antarctica without coming back a different person," Australian palaeontologist John Long says in the introduction to his book *Mountains of Madness*. "A little part of Antarctica grows inside you and moulds your character, for better or worse, from the first day you step foot on that frozen land."

In the summer of 1991–92 Long visited Antarctica for the second time, on a field expedition led by New Zealand palaeontologist Margaret Bradshaw. The purpose of their project was to search a fossil-rich site that Bradshaw had found two years earlier in the Transantarctic Mountains. Their quest: fish fossils from the Devonian Period 400 million years ago. For Long the eight-week trip would induce "a permanent high of either discovery or adventure".

This passage is from a chapter Long called "The first worst day of my life". It is November 1991. Long and Bradshaw, together with sedimentologist Fraka Harmsen and field survival leader Brian Staite, are making their way from the camp at their drop-off point on the Darwin Glacier at 80° South to their field camp in the remote Cook Mountains.

It was a chilly –17°C when we woke up. We finished packing up the camp by 10.30 a.m. and were ready to roll. It was our first attempt in the field at lashing down the sledges tightly and preparing the skidoos for a full day's sledging. We were to head across the Darwin Glacier to the Cook Mountains, which lay to the north of our camp. Yet this was not a simple case of heading in a straight line for the peaks where we wanted to go. Large crevasse fields in the middle of the Darwin Glacier forced us to follow a less direct route, parallel with the glacier, then we had to cut across diagonally towards Gorgon's Head.

Despite its off-putting name, Gorgon's Head is actually a treasure trove of fossils irresistible to the likes of us palaeontologists. The first record of fossil fishes coming from rocks suspected of being southern equivalents to the Aztec Siltstone further north was discovered here a few years before by a young PhD student named Ken Woolfe, now a lecturer at James Cook University in Townsville, Queensland.[2] The fish fossils were only scraps of scales, teeth and bones, but enough of a faunal assemblage was collected to prompt us into writing a small scientific paper outlining the importance of the fauna and recording the presence of Aztec Siltstone in the area.

The sun was shining and the winds were gusting at around thirty knots when we set off sledging across the blue ice that day. Patches of freshly fallen snow were pleasantly smooth to sledge over, but each time we hit the jagged surface of the solid blue ice of the glacier the sledges would lose control as the winds made them drift and sway behind the skidoos.

Stronger winds developed as the day wore on, causing the sledges to blow around even more on the ice, making them difficult to control. Powerful freak gusts would randomly jerk the skidoos and push the sledges close to toppling over. After a few hours of this we stopped to put the metal guide pins down through the sledge runners a notch or two deeper to give more effective grip on the ice surface. It was hard work for the two people on back of the second sledges to steer them over the blue ice during these gusts. Our immediate fear was that the sledges would eventually overturn and crash around on the ice, causing damage to our gear or breakage to the sledge itself, so great care had to be exercised while crossing these wide fields of roughly hewn ice.

As the weather grew fouler I could feel my fingers and toes getting decidedly colder. As it was only our first full day in the field and

---

2 Ken Woolfe died in 1999, just two days after returning from a successful two months with the Cape Roberts Project in the Ross Sea.

I didn't want to seem like a wuss and complain, I said nothing. I was being dragged along behind the sledge in the strong winds and couldn't stop to adjust my clothing, so I had to just keep holding on tightly and working the sledge to try and keep warm. I looked over occasionally to see how Fraka was getting on. She looked rather cold and miserable. At the next stop we rummaged through our packs and donned more layers of clothing. I put on my bear paws. These are thick leather mittens with a fur patch on the back of the hands for wiping the streaming snot from your nose, a common condition that always develops when one is sledging head-on into strong cold winds.

At about 5.00 p.m. that day Fraka was looking seriously cold and couldn't seem to keep warm enough, so Brian and Margaret wrapped her up in her one-piece bunny suit over her already heavily packed layers of clothing. By this stage we had covered a fair distance from the base camp, about twenty kilometres across the glacier, and could see the mountains looming just up ahead. It was imperative to get into the sheltered bays near to the mountains where we might find some respite from the impending storm as soon as possible, so we pressed on relentlessly for another hour or so.

By 6.00 p.m. the storm was gusting fiercely, pushing our sledges around like toys on the glassy ice, even when we were not travelling. We reached the mountains but couldn't see any suitable camp site that could protect us from the roaring winds. Fraka, Margaret and I huddled up behind the back of one sledge, trying to get a solid barrier between us and the biting wind.

Brian then heroically unhitched his skidoo and went off alone towards the mountains, his vehicle being instantly silenced by the deafening roar of the blizzard. We crouched at the back of the sledges and waited for what seemed like an interminably long time but in retrospect was probably·more like ten or fifteen minutes, wondering whether Brian really knew what he was doing. The thought of whether or not we'd ever see him again did cross my mind.

Suddenly, much to our great relief, Brian appeared out of the white blowing snow in front of us. He flashed us a wicked toothy grin saying in a deadpan voice a line I will never forget: 'If you want to live, come with me.' ... We hitched up the sledges and followed Brian into the storm, but about five minutes later we rounded a sharp rocky bluff and found ourselves in a wind shadow. It was still gusting, but nowhere near as fiercely as out on the glacier from where we had just come. Out there we could see huge billows of white powdery snow streaming upwards in geyser-like clouds, changing directions frantically with each erratic wind blast. People would not survive very long out in those conditions, I thought to myself.

The tents were hastily pitched one at a time, with each of us holding down a corner. As soon as they were up we shovelled snow over the flaps, placed rocks on top, and then one person crawled inside to get the cooking gear set up. The others passed in the yellow kitchen box, the green radio box, the blue primus box, and lastly the red food box was placed on the front tent flap close by the entrance. Both tents were soon set up in exactly the same arrangement. We covered up the skidoos with their nylon covers and secured them by anchoring them with tent pegs and guy ropes. Finally, the last job to be done was to secure the twenty-metre-long radio aerial out on bamboo poles in a line perpendicular to Scott Base, with the two thin end wires pushed in through the ventilator pipes at the top of our tent to connect with the radio inside. All of us then quickly disappeared into the shelter of the tents to get warm.

The comforting roar of the primus stoves quickly heated up the pyramidal interior space of the tents. The first thing we did after getting inside the tent was to melt snow on the stove so we could each have a hot drink. We found ourselves ravenously hungry after the day's hard travelling and setting up camp in the strong winds. It had been quite an ordeal for our first full day in the field. Could it get worse than this, I thought to myself, or was this considered to be just an average day in the deep field?

After they had spent two nights sheltering in their tents from gale-force winds and blowing snow, the weather cleared enough for an attempt on Gorgon's Head, a pyramid-shaped mountain 1,960 metres high.

I woke up excitedly next morning at 6.30 a.m., quickly poked my head outside the tent and was relieved to see the weather had improved. I was brimming with enthusiasm at the prospect of getting out on to the mountain to collect fossils. The winds had dropped to twenty knots and the temperature was a cool but tolerable –17°C. We decided it was suitable weather for us to ascend "Gorgon's Head". After a very hearty breakfast we made our scheduled morning radio call to Scott Base, prepared our packs with the necessary gear and food, and set off from base camp at 9.22 a.m. We drove the skidoos about three kilometres from the camp to a low rocky spur which led directly up to the top of the mountain.

It was a long hard slog to the top and took us around five hours, mainly because we were all heavily loaded up with the extra survival equipment and food supplies. The mere thought of finally getting up to some fossiliferous layers had put me in high spirits, along with the rest of the party who were also keen to see some geological action up on the top of the mountain.

On the climb up I collected some fossil plants in the scree fallen from the steep rocky sandstone bluff just below the distinctly coloured Aztec Siltstone layers. On my return to Australia I showed these specimens to fossil plant expert Steven McLoughlin, then at the University of Western Australia, and we wrote up a description of the fossil plants,

A fossil of an ancient bottom-dwelling fish, bothriolepis (right), and the fossil stem of a lycopod, a plant that lived on the continent around 380 million years ago. While lycopods usually grew to less than a metre, some soared upwards of twenty metres.

which was published in the British *Geological Magazine*. The most distinguishing features on the lycopod fossils were the different shapes of the leaf scars on the stems. Lycopods don't have true leaves in the botanical sense; instead they have outgrowths of the bark or stem as leaf-like structures. These eventually detach from the stem as they grow outwards, leaving a characteristic scar shape that can be diamond-like, rectangular or hexagonal, and have a variety of distinctive inner markings as well. To an expert like Steve, the few fossils I'd managed to collect that day contained a wealth of new scientific information.

The specimens of horsetails (lycopods) were identified by Steve as *Haplostigma lineare*, a species otherwise known only from the Late Devonian of Australia; *Malanzania*, a genus known previously only from the Carboniferous deposits of Argentina; and a species of

Archaeosigillaria previously known only from the Devonian deposits of Africa and South America. All of these fossils smacked of Gondwana, reinforcing the important biogeographic links that fossils give for linking the modern-day southern continents.

Staring at the squashed impressions of 380-million-year-old plants while on the top of an Antarctic mountain on a gloomy overcast day brought to mind a very different picture of what this region was like at the time. The land had just been colonised by life for the first time in the Silurian Period, about 420 million years ago. At the time these Antarctic fossil plants were thriving, there would have been a sparse forest of low tree-like plants (the lycophytes), and an extensive ground cover of small stem-like plants (called proteridophytes). These would have flourished near the waterways. The largest of these primitive 'trees' were giant lycopods reaching upwards of twenty metres, but most of the plants grew to less than a metre above the ground. No colourful flowers or pendulous pine cones here, just featureless thin green plants having simple flattened extensions of their stems for leaves.

A few primitive invertebrates, such as springtails, mites, scorpions, millipedes and centipede-like animals, as well as early spiders and their close cousins the trigonotarbids, crawled around these forests. They fed largely on each other as herbivory hadn't yet been invented. Creatures just didn't have the specialised ability to digest cellulose in those days, so they just had to be content to eat each other. On the land the scene was essentially quiet, but the real action in this Devonian world was taking place below the water, in the large flowing rivers and expansive lakes, which teemed with many kinds of primitive fishes. Some were the size of tiny minnows; others were huge predatory monsters four metres long with razor-sharp stabbing teeth six centimetres in length. This was the real reason I had come to Antarctica, to collect the fossil remains of these most interesting ancient fishes from a time when fishes were the highest form of life on Earth.

Sedimentologist Fraka Harmsen at the crater of Gorgon's Head, summer, 1991.

We reached the base of the Aztec Siltstone, near the summit, at around ten minutes past three that afternoon. Finally I was face to face with the variegated green, red and grey banded rock unit famous for its ancient fossil fish remains. After about three hours of searching I located a fish fossil horizon about twenty metres from the top of the Aztec Siltstone, more or less at the same level where the original fossil fish material had been found by Ken Woolfe's party three years earlier. The fossil bones showed up as light specks and streaks in a dark-green silty sandstone. These layers of rock did not outcrop very extensively and were spasmodically covered by the loose scree. I could trace the fossil-bearing layers only as small lenses of about two or three metres each in length. Further up the hill near the top of the Gorgon's Head we located another two fish-bearing horizons and proceeded to spend an enjoyable afternoon collecting lots of specimens. Most of these were small fragments of bone or teeth, nothing of spectacular appearance to the layperson, but many represented new records or unknown species for this region, so I was extremely happy with the things we were finding. Often this work required sitting down with a pile of rocks and carefully examining each surface with a hand lens to find the really interesting small things, such as the beautifully preserved teeth of sharks, or exquisite fish scales adorned with complex sculpturing on their surface.

As I worked on hands and knees with my nose down on the rocks, Margaret and Fraka measured a detailed section through the Aztec Siltstone, aided by Brian, who held the measuring staff and lugged gear around for us. The sky was an ominous dark grey, reminding me of the gloomy Melbourne winters of my childhood, but as there was no wind or snow we were able to keep working up on the top of the mountain until about 8.00 p.m. Then, hitching our full backpacks bursting with specimens, we turned our back on Gorgon's Head and started heading back to camp.

From *Mountains of Madness* by John Long: Allen and Unwin, Sydney, 2000.

Scientist Bill Green became fascinated with the landlocked lakes of the McMurdo Dry Valleys. How had these bodies of water evolved? Why were they so different in their chemical composition? In this photo of his camp at Lake Hoare are two of the young scientists who took part in his research.

Water, ice and stone

"When I tell people there are lakes in Antarctica, they think surely I am joking." American limnologist and geochemist Bill Green was referring to the lakes of the McMurdo Dry Valleys. These landlocked lakes were, he said, "without spillage or outflow; each has only a few streams, and these hold water for only a few weeks out of the year. They are ice-covered, so that very little in the form of dust or snow enters them from the air. And, of course, there is never rain. That in themselves makes them magic."

Green's book *Water, Ice & Stone* draws on his research in the Dry Valleys through the 1980s and '90s. His ten-year investigation sought answers to questions such as how and when had these odd water bodies evolved? Why were they so different in their chemical compositions? How were they sustained biologically in such a harsh environment? How were the chemical elements transported to them, and how were those elements removed?

He wrote the book in a desire to share how it felt "to do field research in the hostile, austere but beautiful environment of the Antarctic continent". In this passage he describes a day's fieldwork at Lake Miers, in the Miers Valley, where he was working with an engineer colleague, Larry Varner, and a team of young scientists and what they saw beneath the lake's frozen surface.

After breakfast, Mike and Walt finished stringing the antenna wire across the fronts of the four tents. They had tied one end to an upright shovel and the other to a six-foot length of drill extension. The antenna was nearly fifteen yards long, and it paralleled the ground at a height of five feet. Walt was kneeling beside it, holding the dark handset of the radio.

The bulky leather mitten of his left hand was pressed up beneath the hood of his parka, and his head was bent down into the falling snow as though he were trying to envelop whatever sound might eventually come forth. His voice was getting louder and more deliberate as he enunciated, "Mac Sideband, Mac Sideband, this is Sierra Zero Four One. Do you copy? Over."

From the headset there came an immediate burst of garbled sound like rocks rolling over sheet metal. He reached for the clarifier and adjusted it, but the sound remained harsh and unintelligible, as though it were a stream of "r"s and "t"s roaring and hissing at him out of the cosmos. Not speech. I knew he would be kneeling there for a long time.

Varner sat on a trunk with a journal opened on his lap and a pencil poised above the rustling pages. He looked out across the brittle snow patches down toward the lake, and then back up the valley to the glaciers. He lowered his eyes and began to write. Mike and I loaded the sleds for a full day of sampling. We packed up the Kemmerer, the barrel filter, two dozen polyethylene bottles, a Secchi disk, the oxygen and pH meters, nitric acid and Eppendorf pipettes, and some cheeses and bread for lunch. Walt, having failed to reach Mac Sideband, patiently turned off the battery-powered radio, attached to it a small array of photovoltaic cells that he opened wide to the sunless sky, and walked over to join us. By now Varner, who had stopped writing, decided not to risk the slippery lake surface with his bad knee, and was already walking pensively up toward the Miers and Adams Glaciers. He was clutching the notebook and pencil in his gloved hand.

It was about ten o'clock when we walked on to the surface of the lake, but time had become a useless convention, an artifact of monasteries and corporations, a trick to divide the seamless day. Here, with constant daylight, nothing rang the hour. What time was it really? What aeon?

We took the same path Walt and I had followed yesterday. The sleds trailed noiselessly behind us over the smooth ice with its dusting of fresh snow. The snow itself seemed hardly to be falling at all, seemed to be hanging suspended as in a paperweight. As we turned in toward the hole, the sleds began to shudder over the rough ice. Between two cairns of heaped stone we could see our red and green marker flags hoisted on bamboo above the hole we had drilled. We were nearly at the sampling site when Mike finally exclaimed, as if doing a double-take, "What! Why are these stones here?"

The stones were astonishing. They lay in heaps and cones across the surface of the ice, usually at a considerable distance from the shore. There were boulders mixed with pebbles and gravels and wind-blown sands, and in the centre of each pile was a core of dark ice that stood protected from the winds. Ice-cored drift. Walt had suggested yesterday that the boulders might have rolled down out of the mountains on to the lake surface. But when he looked around at the slopes he reconsidered and realised they were far too shallow and the boulders far too irregular for this to have happened. "Bowling balls might have made it down on to the ice," he had said with a grin. "Boulders never." These were like the monoliths of Easter Island and we accorded them the same respect born of mystery.

The level of the water in the hole was a few inches below the ice surface. The hole had frozen over last night, and we had not brought the drill with us. Mike got down on his hands and knees, reached in, and hit the ice with his fists and with the palm of his hand. But it was more than an inch thick and wouldn't crack. Walt walked over to the piled drift and picked out a sharp boulder that he carried back in both hands. From a standing position, he threw it full force into the hole. Where it struck you could see a frosted indentation that looked granular, like sugar. Outward from the pulverised centre, little cracks radiated away toward the edges. Mike reached in again, took the rock, and beat against the surface until it collapsed with that whisper that new ice sometimes makes. Freezing water rushed in around his hands and he quickly pulled them out, dropped the stone, and thrust his hands into his coat pockets.

Walt unravelled the nylon line that he had marked off so meticulously back at McMurdo. At one-metre intervals, beginning with the midpoint of the sampling bottle, he had wrapped small pieces of waterproof yellow tape around the line. At five, ten, fifteen metres and so on, he had added a red band; and at each ten-metre interval he had placed an additional orange stripe. When the first tricolour of tape crossed the plane of the hole, we would know that the sampler lay ten metres below. I scooped ice from the hole and peered down into the depths of the lake. I was looking through a crystalline pipe whose sides appeared undulant and glassy and tinged with blue. Waves of blue glass. Below it, all around, falling away into an azure depth, lay the clearest water I had ever seen.

Walt opened the gray barrel of the Kemmerer bottle at both ends so that water could pass through as it sank. The bottle had a large volume, more than six litres, and was constructed entirely of plastic. With his feet spread so that he stood directly over the hole, he lowered the sampler until its bottom end touched the water's surface. He let it sink slowly, let the water rise up gradually through the cylinder until it finally covered the top.

He fed more line into the lake, watched the yellow stripes of tape slip beneath the surface. When the sampler had fallen five metres he stopped it, squeezed the line between his thumb and forefinger, let it hang there for a while, still perfectly visible, moving back and forth like a massive chandelier. I took the "messenger" from the box, clipped this simple metal bar on to the line, and with a flick of the wrist shot it down into the lake. When it hit the Kemmerer, the bottle closed instantly and huge bubbles, like translucent fish, welled to the surface. Hand over hand, Walt brought the sampler in, held it upright on the ice, and pried open the top. It was filled to overflowing.

An aerial photograph of Lake Miers and, in the background, the
Miers Glacier.

We sampled for twelve hours. The open Kemmerer slipped into the lake, submerged itself with a "Glub!" as the frigid waters poured over its lip, down the inner sides, and back into the lake. It fell along the crystalline ice, touched as it sank, gently bounced away, centred itself in the hole. We lowered the sampler, checked the coloured markings against the ice, double-checked, sent the messenger, waited and waited and felt for the tug, the tug of the filled sampler stretching the line. We pulled it up as water from the line beaded into tiny droplets, filled the air, and froze as we worked. Into the polyethylene bottles arrayed along the sled we emptied filtered water, six litres from each depth: two litres for nutrients; two for metals; two for major ions. We labelled and relabelled the bottles, wrote in waterproof pen, assigned each a number, and in one of the little green notebooks wrote what that number meant, where the sample had come from, and on what day. By evening our gloves had become blocks of ice.

Toward McMurdo Sound, beyond the drift mounds, the sky had turned light. Opalescent clouds, suffused with wisps and streamers of fine-spun pink, stretched above the sea ice across the broad opening of the valley. In the middle of this, just where the curve of the mountains reached its lowest point, stood the hills of an island, visible for the first time. The sky to the east was warm and light-filled as though the sun had just set. For a moment it made me think of evenings on wooden decks, of olives and palm leaves, the flatness of a summer sea.

We loaded the sleds, stowed the tightly capped samples under the heavy canvas, and tied them in. There were bits of plastic and Styrofoam chips and pieces of string and Tygon strewn around the hole or refrozen into small lenses in the ice. We pried them loose and picked them up as best we could, until the surface looked much as we had found it.

Before we returned to camp I wanted to take one more measurement. Out of the prow of the sled I pulled a disk the size of a pie pan. The Secchi disk was forty centimetres in diameter and divided into alternating black and white quadrants. Beneath the disk a small square was attached as a sinker. On top there was a ring to which a line could be clipped. The idea was to lower the disk into the water until it could no longer be seen. In the limnologist's trade, among "the gear and tackle and trim" this was the simplest of all things – a circle in search of the dusk, in search of that thin stratum of lake where light and dark merged to become one.

Though over the years it has become one of the essential tools of limnology, the Secchi disk was named in honour of an astronomer. In the nineteenth century, the wells of Italy became contaminated. The contamination, it seemed, was related to the turbidity of the water source. The Pope, fearing an epidemic, called upon the famous astronomer and populariser of science, the priest Angelo Secchi, to devise a reliable means for determining the comparative clarity of well water. Secchi had already pioneered the use of photography as a tool in astronomy, and he had been among the first to obtain spectra of the planets, the sun, and the brighter stars. A master of telescopes and spectroscopes, he was no stranger to the ways of light. Father Secchi accepted this hydrologic challenge. He fashioned a small disk that could be suspended from the end of a rope and that could, within seconds, provide information about the quality and safety of a water supply.

Secchi's disk is used today to determine water clarity, and limnologists have obtained "Secchi depths" on lakes throughout the world. Not surprisingly, there is a reasonably good correlation between the Secchi depth of a lake and its biological productivity; in general, the correlation says, the more biologically productive a lake is, the more organisms it supports, the more light it scatters, the smaller will be its Secchi depth.

As the weather cleared and the sky became a powdery blue over the eastern valley, I began to lower the disk through the hole. It settled into the water, jerking a little from side to side as it sank. I could see it clearly as the five-metre stripes passed over the ice, and I could still see its black and white quadrants as the first tricolour touched the water. At fifteen metres it appeared small and far away, like a silver coin dangling on the end of a string. In another metre I had lost it. "Sixteen metres," I called to Mike, and he wrote it down in his notebook. The lake was clear to more than fifty feet. ...

We moved the sleds out, turned them west toward camp. Blue sky shot in streaks beyond the glaciers. The lake ice, in spots, had become like glass, and its sculptures bent the rays of evening light. As we pulled I thought of the messenger: the way it fell along the line; the way it shook the silver bubbles from the lake, shook them into bursting like grapes. And how it closed the sampler around a precise parcel of water – the one that we would come to know in our limited way as the lake itself.

I thought of all the messengers we sent into the folds and fires and distances of the world, into seas and galaxies, into its microscopic hollows and sockets, its very being. Those messengers of our craft and invention returned to us from those places – bights and hinterlands below our seeing, beyond our imagining, pasts of warmth and abundance – returned as emissaries bearing gifts and tidings. Alpha particles transiting the disks of atoms; ionic currents streaming in laboratory glass; starlight and planet-light spreading like fingers through a prism in a dark room; the ladened ships of exploration returning, struggling to return; all the imprints and splinters and shards of bone pressed into ancient rock; heat, pulsed into a cup of ice; stones heaped quietly upon a lake. All of them – light waves, water waves, particles and shells, wisps of energy and reflected sound – all were messengers. The disk itself was a messenger, and the Kemmerer, and the metres all seeking the measure and structure and order of the world.

From *Water, Ice & Stone: Science and Memory on the Antarctic Lakes* by Bill Green: Bellevue Literary Press, New York, 2008.

The lakes of Mars

Not airless,
but there's no insect life
to set the air moving;
no birds this far inland,
and even the mummified seal,
lying not far from the shore,
found her spacesuit lacking
as she crawled here from the coast
a hundred years ago:
forty miles inland, she starved
and crumbles at her own pace,
friable witness among the stones
to the exobiology of this
open-air refrigerator.

It's business as usual:
we hike across Lake Bonney's
star-fractured lens,
taking ourselves further out
than we've ever gone before,
right to the planetary edge
and beyond, to what this lake
neatly simulates: an ice-bound
port to the solar system,
a mirror of the lakes of Mars.

Crunching over waters
that carry the imprint of BC
in their toxic depths, we are
as tentative as Gulliver
among miniature cathedrals of ice
shattering at our approach.

Far far below the two-metre
plug of ice, below the spoked
and fairy-palace shimmer
of the lake surface, there's
the murmur of underground cisterns,
a tidal edginess or eagerness
in the buried water, a glitch
that erupts now and then
like static.
          We carry a thick plank
to bridge the slushy shore ice;
hoisted on a shoulder, it's acquired
an absurd value here where timber
is scarce. The sun this morning
burns as cool as a Davy lamp,
and a thin layer of air above
the surface comes at us like a sheet
shaken out on a frosty morning.

One part of the lake is fanned out
in icy gravestones, yet there is
no company of the dead
anywhere near this valley,
and menace is defused in the scene
by our low jesting banter.

There's just that note of exile,
grey dregs of a moonlit walk
in a dream or nightmare,
as the hinterland looms
across the far shore, and
Charon's boat is a raft of ice
fixed fast to its moorings.

Someone
bends down and eats some ice:
It's bitter, he says, *tastes like iron*.

Chris Orsman

Katabatic winds at Cape Denison, which Douglas Mawson described as "the windiest place on Earth". In this photo, two members of his Australasian Antarctic Expedition, Leslie Whetter and John Close, are trying to collect ice from a glacier adjacent to the huts.

Katabatic winds

While working on a degree in geology and geophysics at Rice University in Houston, Stephanie Shipp collaborated with the Education Development Center and the American Museum of Natural History to develop school curriculum material about Antarctica. This blog post, one in a series called "Letters from Stephanie", describes and explains the katabatic winds that dominate in parts of Antarctica. The word "katabatic" derives from the Greek "katabatikos", meaning to flow downhill. Shipp is on a drilling ship stationed off the coast of East Antarctica.

I'm writing to you under some pretty tough conditions. We hit a nasty windstorm today. Winds are reaching 80 knots – that's 92 miles per hour. The waves are kicking up and splashing over the side, half our crew is seasick, and – well, let's just say we haven't gotten much work done.

This storm isn't exactly a surprise. We are trying to collect some sediment samples and sea floor images from the ocean floor around Cape Denison, which is known as the windiest spot in Antarctica. And Antarctica itself is the windiest place on Earth. In fact, Antarctica holds the record among continents for sustained wind speeds; wind speeds can reach 200 miles per hour here. Most research teams, like ours, learn to deal with the wind; other teams actually spend their days in Antarctica studying wind.

The winds of Antarctica, however, are a tough study. Even in places where winds are a little less extreme, the wind often damages the weather stations used to measure it. The wind blows snow into and out of precipitation gauges. It kicks up blinding blizzards that keep cargo and personnel transports prisoner. High winds and blizzards keep research teams holed up in their tents for hours or even days, unable to venture into the field.

Why are the winds of Antarctica so powerful? Antarctica's temperature patterns are part of the reason. On warmer continents, air temperature usually decreases as distance above land increases; in other words, the higher you get, the colder the air. When you climb a hill or a mountain, it gets colder toward the top, right? But in Antarctica it's a bit different. Antarctica's interior region is the high Polar Plateau, an area covered with a thick ice sheet. This massive ice sheet cools the air above it; as a result, all the air above the Polar Plateau is very cold. The coldest, most dense air is closest to the ground; the higher the air, the farther it is from the cold ice sheet. That means that air temperature rises (instead of decreasing) with distance above land. This is called a temperature inversion because it is the inverse, or opposite, of more common temperature patterns on warmer continents.

Cold air is denser, or heavier, than warm air; that's why warm air rises and cold air sinks. (Think of how hot an attic can get in the summer.) The Polar Plateau is covered with so much ice it is always cold. This means it is always cooling the air above it. As a result, a mass of very cold dense air sits on top of the plateau. This cold dense air wants to sink, so it flows down from the high continental interior toward the lower coast, just as a stream flows down a mountain. These interior winds of cold dense air are called inversion winds.

As you can see, most of the inversion winds flow fairly evenly down a gentle slope. However, some of the landscape is not a gentle slope: winds can be channelled by the rugged landforms of ice and mountains. Picture liquid being poured into a funnel – when the airflow of interior winds converges, more air is being compressed into a smaller channel space. The airflow gets stronger, turning into fast-flowing winds called katabatic winds. These katabatic winds roar toward the coast of Antarctica. Fairly quiet conditions turn instantaneously, with katabatic winds reaching speeds of fifteen to twenty metres (fifty to sixty-six feet) per second.

Things are getting pretty wild here on board; I had better go and help everyone lash things down. We try to stow our equipment and samples safely at all times, just in case we run into heavy seas. But there are always a few things we have overlooked that we need to tie down – microscopes, cabinet doors, computers, my shift-mates …

From "Katabatic Winds" by Stephanie Shipp in *Curriculum Collections*: American Museum of Natural History, amnh.org.

Emperor penguin chicks close to fledging. The emperor penguin colony close to Halley Research Station on the Brunt Ice Shelf is present from May until February.

Inside the emperor penguin egg

In 2003–04 Gavin Francis spent fourteen months as camp
doctor at Halley Research Station, the British Antarctic Survey
base on Brunt Ice Shelf in East Antarctica. His routine work
included training staff in first aid, monitoring dental and
mental health and treating minor injuries, but he had to be
prepared for any possible accidents or medical emergencies:
Halley is accessible by ship for only two months a year and is
so remote it's said to be easier to evacuate a medical casualty
from the International Space Station than to bring someone
out of Halley in winter.

Like many of the early Antarctic expedition doctors, Francis
also used his time on the ice to indulge his passion for natural
history in general and emperor penguins in particular. From
Halley it was a twenty-kilometre snowmobile journey to the
nearest rookery, where 60,000 emperors bred each autumn. On
one of his summer visits, Francis collected some abandoned
eggs and kept them in a freezer at Halley until he was ready
to investigate.

I felt anxious about breaking into the first egg. I had no idea if the
penguin chick within would still be microscopic, too small to see, or
almost fully formed and difficult to get out. Even though any chick
inside would have been dead for nearly a year there was a trepidation
about cracking the shell that had something to do with reverence for
life, and something to do with respecting the dead. It was early June;
down at the coast the emperors would be laying. Thinking about the
eggs and Wilson, Cherry-Garrard and Stonehouse, I had wrapped my
own stash of emperor eggs in a plastic bag and thawed them out. After
a few moments' deliberation I cut a hole in the base of the first one
with a drill and peered in.

A tuft of black feathers surfaced. I widened the hole, and using some tweezers managed to get hold of the chick's beak. I drew it out and, opening the hole further, delivered the crown. Its head swung down, neck spindle-thin, like a suicide on a rope. Its eyes were closed; black eyes with white patches around them, a panda cub in negative. Emperor penguin chicks have an egg tooth, and a special neck muscle which withers after hatching. Even so, it can take them three days to break through the armoured shell. Perhaps this one had tried to do so. I cracked the rest of the egg and admired the neatness with which the chick fitted, the tapering of the tail and hindquarters into the apex of the shell, the economy of nature. I had found some stainless-steel trays in the medical room and laid it out like a body in a morgue.

The next egg was easier. I punctured the shell and the albumen started to spill out, then stopped abruptly. Looking inside I saw a fleshy knob was blocking the hole, rounded like a fingertip in a dyke. I nosed it up and again caught the beak in the tweezers, delicate as a twig, and pulled it through. Its naked neck was ribbed like an earthworm and its skin was pale, almost translucent. It was rubbery and etiolated like grass left under a stone. It looked mid-stage, maybe a month's gestation. Through the thin skin of the head I could see the dark outline of eyes that filled the skull, eyes that would, if they had survived, have tracked the glint of fish half a kilometre under the sea. When Wilson collected eggs on those earliest *Discovery* trips to Cape Crozier he had hoped to find an embryo at just such a stage as this but all of his eggs had been much further developed.

The rest of the body slipped out, followed by the yolk sac and the remainder of the albumen. Its wings and feet were stubby and translucent, like tadpole tails, and pinprick feathers shadowed the skin around the abdomen and tail. They rubbed away in my gloved hands.

A third egg was addled, a turbid soup, coloured and fronded like Sargasso weed. I flushed it away for its sulphurous stink. A fourth was well preserved but the embryo inside, a dark germ on the yolk, was too early to see properly. In the medical room I had an ancient microscope for performing blood counts on humans but none of the chemicals I would need to fix dead penguin tissue. I realised that I was wholly unprepared for carrying out any serious or helpful research on the emperors, alive or dead, though there is still a great deal unknown about them that I could perhaps have worked on. ...

Despite my lack of equipment or direction I got on with studying them in my own quiet way, not contributing anything to the world's knowledge but attending to my own curiosity, sense of wonder, and gratitude that as human beings we were not alone in this place.

From *Empire Antarctica: Ice, Silence and Emperor Penguins* by Gavin Francis: Chatto & Windus, London, 2012 and Counterpoint, Berkeley, 2013.

Twin Otter aircraft, equipped for polar conditions, were used to search for lakes and a 1,200-kilometre-long mountain range buried deep beneath the ice of East Antarctica. Ice-penetrating radar antennas and magnetometer pods are mounted on to the wings.

The mountains under the ice

In 2004 a team led by American marine geophysicist Robin
Bell discovered four subglacial lakes buried deep beneath the
East Antarctic Ice Sheet at the foothills of the Gamburtsev
Mountains, a massive range buried deep beneath the ice. In
2008, as part of International Polar Year (IPY), Bell led an
expedition that involved using remote sensing technologies
to explore the mysterious mountain range and lakes.

The Gamburtsev Mountains were discovered in 1958 by
Soviet geophysical surveyors, but the presence of such a
large range in East Antarctica did not fit with scientists'
understanding of the continent's geological history. As Bell
put it, the discovery was "equivalent to an archaeologist
finding a fully suited astronaut inside a pyramid".

Understanding the origin of the mountains was the primary
goal of the seven-nation IPY project and it explored the
subglacial landscape using two Twin Otter aircraft equipped
with ice-penetrating radar, gravity meters and magnetometers.
An additional goal was to gain a better understanding of
Antarctic glacial plumbing. Bell hoped to investigate why
the subglacial lakes were there, if they were changing, and
to what extent the lake system was influencing the subglacial
flow of polar ice toward the oceans.

Before heading to Antarctica, she wrote this feature for
Scientific American, which she introduced by talking about
the North American lakes with which she was most familiar.

When you walk through the woods toward a lake, first you will often
see ducks, loons, mergansers or other waterfowl flying intently towards
the water. Soon you may notice marshy wetlands. Getting closer you
will hear the harsh rattling call of the kingfisher waiting to skewer a

fish from the lake. As the trees thin, the horizon will open and the waters of the lake will stretch before you.

Walking toward a subglacial lake there are far fewer clues. There are no trees to obscure your vision, only white snow and blue sky in every direction. The only hint of the lake will be when your colleague walking a quarter of a mile ahead suddenly disappears into a fifteen-foot (4.6-metre) deep moat. This 2.5-mile (four-kilometre) wide moat is the result of the ice sheet "sagging" as it goes afloat over the lake. There will be few other clues of the lake beneath your feet. The two miles (3.2 kilometres) of ice effectively hide the underlying terrain and the winged waterfowl have long ago fled these Antarctic lakes.

Appreciating a subglacial lake requires a little more distance to get perspective on the vast, apparently featureless ice surface. The first person to get a little distance was a Russian pilot who transported scientists and engineers between the Soviet camps in East Antarctica in the 1950s. Staring out the window when the sun was low, he began to notice that there were large, extraordinarily flat places in the ice sheet. He catalogued these sites on the aviation maps with the hopes of writing a thesis on the phenomena upon his return to Moscow. Unfortunately, he was killed in a crash and subglacial lakes were relegated to myth status in the polar community.

It was almost as if subglacial lakes were destined to remain hidden. Early seismic data was misinterpreted. Records were destroyed in fires. Equipment failed just as teams driving hundreds of miles reached the edge of the lakes. Some tantalising hints suggested there might be water under the ice sheet, but it was not until we had the perspective of looking at the ice sheet from space that the large lakes became evident. Detailed measurements of the height of the ice surface provided the first real opportunity to "see" the lakes. The surface of most of the ice on the sheet is rough as it flows over the hills and mountains below.

Just as in winter a woodland lake will be an expanse of horizontal floating ice continuing to the horizon, the ice above a subglacial lake floats and the ice surface is very flat.

These lakes exist because the thick ice acts as an insulating blanket, capturing the heat emerging from the Earth's interior. The temperature at the top of an ice sheet is minus 50 degrees Fahrenheit (–45.5 degrees Celsius), while at the bottom it is positively warm at 28.4 degrees F (–2 degrees C), very close to the melting point of ice. The pressure of the overlying ice lowers the melting temperature of ice a bit, but the main reason for the warm temperatures is ... the natural geothermal gradient.

Twenty-five years ago no one would have believed there could be lakes under the ice. Ten years ago scientists thought these lakes were stagnant and isolated from one another. Today we know that sub-glacial lakes are connected under the ice through a maze of plumbing, and that this connectivity can subject them to rapid drainage, akin to pulling the plug from a bathtub, allowing water to drain from one lake into another. Draining water from subglacial lakes may contribute to the onset of ice streams, accelerating their movement toward the continental edges, where they rest against the surrounding ocean water.

In East Antarctica the largest subglacial lakes are found in the foothills of the Gamburtsev Mountains. Lake Vostok, the size of Lake Ontario, and two other large deep lakes mark the eastern edge of the mountainous province. On the western edge, four large lakes are linked to the onset of the rapid flow of the ice sheet. ...

If the East Antarctic Ice Sheet were dropped on top of the lower forty-eight U.S. states, every single town would be covered. Only a few mountain peaks would be exposed. Satellites cannot see through the ice sheets. Studying mountains and lakes covered by a thick blanket .of ice is a challenge.

Fifty years ago, scientists had no good estimate of the thickness of the Antarctic ice sheet. At the beginning of the last International Polar Year, using oil industry technology, convoys of tracked snow vehicles from many nations set out across Antarctica. The convoys, or traverses, stopped every fifty miles (eighty kilometres) to lay sensitive recording devices, geophones, drill a 150-foot (45-metre) hole, and set off small explosive charges. The explosions would send a fountain of snow into the air and energy deep into the ice sheet. The downward propagating wave would bounce off the hills and valleys at the bottom. The return echo would be recorded by the geophone. Each explosion produced one measurement of ice thickness.

Despite the very slow progress of these surface vehicles, one of the major discoveries from 1957 International Geophysical Year resulted from the recordings of these explosions. The scientists, bundled in parkas, had to wait much longer than anticipated to record the echo from the ice-sheet bottom. The reason: the ice was thicker than predicted. The East Antarctic Ice Sheet is up to 2.8 miles (4.5 kilometres) thick in places, enough ice to raise sea level globally 170 feet (52 metres) if it were to melt.

Driving over the ice sheet and setting off explosives every fifty miles is a slow process. Flying, even in a small airplane, is much faster. We will use two aircraft bristling with antennas and stuffed with instruments to collect new measurements of the ice sheet from the air, along with those taken from seismometers buried in the snow.

Mounted on the wings of the aircraft are eight antennas that transmit and receive 150 megahertz pulses to measure ice thickness. This radar system, developed by the Center for Remote Sensing of Ice Sheets in Lawrence, Kansas, has been developed specifically to image through the polar ice sheet. Similar to the seismic method, energy is transmitted through four of the antennas. The energy bounces back from the surface of the ice and the bottom of the ice sheet. We end up with thousands of measurements every second. This is a big improvement over fifty years ago when it took two to three days to measure ice thickness.

To get a better idea of the surface of the ice sheet, my colleague Michael Studinger has installed near-infrared laser, mounted below the floor in the aircraft. Within the nitrogen-filled container the laser fires at a revolving mirror. When the mirror spins, the laser is aimed first to one side of the aircraft then to the other, so that the laser measures the ice surface right below the aircraft and out to the side. When we accurately position the aircraft, the laser measures the distance to within two inches (five centimetres). ...

Along with the ice thickness data, we want to understand the origin of the Gamburtsev Mountains. To do this, we need to decode the fundamental structure of the crust and lithosphere beneath. It will be years, maybe decades, before anyone drills into the Gamburtsev Mountains so we will use gravity, magnetics and seismic velocities to remotely probe the subglacial terrains. The Earth's gravity field changes depending on the type of bedrock. A stronger gravity field means denser rock and a weaker gravity field means a less dense rock. An extremely accurate gravity meter will be mounted in the front of the aircraft. Measurements of variations in Earth's magnetic field will tell us about the nature of the underlying rocks. Some rocks are much more strongly magnetised than others. We will measure the changing magnetic field with caesium-based sensors mounted on the tip of the aircraft wings. Measurements of the magnetic field will tell us how magnetic the hidden rocks are.

The laser will measure the ice surface. The radar will measure the hidden topography. The gravity and magnetics will tell us about the make-up of the shallow part of the Earth – in general the crust. If we want to see deeper we need a different method. A team lead by Doug Wiens from Washington University in St Louis will install twenty-six seismometers spread hundreds of miles apart over the top of the Gamburtsev Mountains. ... These seismometers will be left in place for a year buried in the snow; they will be powered by the sun in the summer and by buried batteries through the long polar winter. The seismometers will record earthquakes from around the world to map the deeper structure beneath the mountain range. In the end the seismic data will let us know how fast the energy from the global earthquakes travels and how warm or cold the Antarctic continent is at depths of hundreds of miles down.

Using this full suite of geophysical techniques, we will try within the next three months to build the first comprehensive cross-section of the largest ice sheet on our planet. In the world of modern imaging, we are accustomed to seeing X-rays of our teeth, MRIs of our brains, and even the fuzzy images of fetuses in utero. These technologies have advanced far enough that now we are not surprised to see medical imaging systems installed inside Winnebagos parked alongside our local malls. Medical technologies have to image through tissue and bones. Imaging through two miles of ice is similar but requires different strategies. The Gamburtsev Mountain expedition is our best chance to image and to understand the ice sheet and the mountain range.

While the peaks of the 1,200-kilometre-long mountain range were found to reach elevations of up to 3,000 metres – about as high as the European Alps – the mountains were hidden beneath 600 metres of ice and snow. One of the surprising discoveries was the age of the mountains: there was evidence the range first emerged one billion years ago, and that eroded remnants of the mountains were uplifted again between 250 and 100 million years ago.

The next goal of the team is to drill through the ice to collect rock samples from the Gamburtsevs. "Amazingly, we have samples of the moon but none of the Gamburtsevs,"

Bell said in 2011. "With these rock samples we will be able to constrain when this ancient piece of crust was rejuvenated and grew to a magnificent mountain range."

From Robin Bell, *Dispatches from the Bottom of the Earth: An Antarctic Expedition in Search of Lost Mountains Encased in Ice*: scientificamerican.com, November 12, 2008.

Frozen Lake Untersee in Queen Maud Land. For Michael Becker, diving under the ice was "like being trapped beneath an unbreakable ceiling with only one small window of escape".

Antarctic time capsule

Michael Becker, a doctoral student at Canada's McGill University, was a scientific diver on a 2103 expedition to Lake Untersee in the Queen Maud region of East Antarctica. Scientific divers first entered the lake in the 2008–09 season as part of a search for extremophiles, life forms that have adapted to extreme environmental conditions such as hot springs. sub-zero lakes and glacial ice. Here Becker heads beneath the ice.

After much anticipation and with a dash of apprehension I dropped below ten feet of ice cover and into a place cut off from all of human history. This is what the unknown feels like.

I had run the scenario of what the dive would be like for days and weeks. I had read stories and talked to the people who had done it. But there is still a great deal of trepidation before you take that final drop into the water and descend into an ice-covered world, totally alone.

After weeks to think about the dive, it both confirmed some of my expectations and proved wrong many of my assumptions. The first was that it would be unbearably cold. The water temperature is 32.5 degrees Fahrenheit and on long dives you can even see ice refreezing at the top of the dive hole. But throughout the dive I was surprisingly warm. It may have been the excitement or the distraction of all the new stimuli – it is a lot to take in at once. But despite the comforting warmth, diving under the ice is like being trapped beneath an unbreakable ceiling with only one small window of escape. If your gear breaks or you run out of air there is no fast exit – you must swim as quickly and calmly as you can toward that glimmering four-foot-wide beacon of hope.

I was a bit concerned about this before getting into the water: you can never be totally sure your gear won't fail. And with such low water temperatures there is a real danger of a freeflow on any dive. That is when the scuba regulator that you breathe through gets stuck in the "open" position, and the contents of your air tank come racing out uncontrolled. This leaves you minutes to get to the surface and get your head above water before you run out of air.

With this in mind some prior apprehension was warranted, but during the dive I felt much more relaxed about the endless expanse of ice above me. Looking in any direction all you see is deep blue abutting a subdued white. The ceiling is marked with innumerable gas bubbles, some so large they look like giant lakes of air trapped in the water. They shimmer and melt together like oversized pools of mercury. Swimming away from the dive hole I was able to put the nerves to rest, but I always kept one eye on that shining dive hole in the distance.

I had imagined being under the ice would be a much darker, spookier affair: ice cover absorbs a lot of the light that would otherwise illuminate the water, and light is attenuated on any dive at depth. But this lake had the same color at forty feet as the ocean has at 130, and life just below the lake's ice visor was more a surreal blue than an inky dark. There was no need for a flashlight, and indeed once my eyes adjusted it felt quite natural. And that was the biggest surprise of all – how calming it felt.

It is this calm, this seclusion, that has allowed cyanobacteria, one of the oldest forms of life, to persist and flourish down here. They are the reason we have come here. The microbial communities in Lake Untersee are so dense they form microbial mats and structures large enough to see with the naked eye.

Cyanobacteria are resourceful organisms. They produce their own energy using a photosynthesis process similar to plants. Light penetrates through the ice cover and through the water, down to the lake floor where the microbes grow. Because of a lack of predation and a lack of disturbance by larger organisms, these microbes grow rampant over any surface that is hospitable for growth – in this case any depth that light can reach.

This type of microbial growth is actually quite common in Antarctic and Arctic lakes. What makes Lake Untersee particularly special is that these microbes – collections of millions of individuals growing together over thousands of years of layered development – form two different types of structures: pinnacles and cones. Both these types of microbial communities represent one of the earliest forms of life, present in the fossil record nearly 3.5 billion years ago.

The pinnacles of our lake are small, between half an inch and six inches tall, and are dominated by a *Leptolyngbya* species that is common in Antarctica. The other structures, found nowhere else on present-day Earth, are the conical stromatolites. The microbes covering the cones are not very thick, roughly half a millimetre, but the dark purple cones can grow higher than one and a half feet. These cones consist predominately of a *Phormidium* species, and it is thought that the different microbial species are somehow responsible for the creation of the different structures.

While it is not known how it happens, these different structures are probably a result of different, community-specific responses to the environment. The fact these two structures grow next to each other tells us that the different microbial mats are responding to the same lake environment in very different ways. It's not clear why or how these unique cones are formed in this lake. Indeed, this is the major driver of the research project here and what makes this lake particularly special.

These microbial mats once dominated the Earth. As multicellular life developed, these early life forms were outcompeted and restricted to fewer and fewer places. Today, there are a handful of rare environments where they can flourish undisturbed. It is here, at the bottom of the world, that we can look through a window into our very distant past.

From "Diving an Antarctic Time Capsule Filled With Primordial Life" by Michael S. Becker, in scientistatwork.blogs.nytimes.com, January 31, 2013.

Scientific instrument ANITA III, ready to be launched from McMurdo
Station to scan the Antarctic continent for evidence of neutrinos and
cosmic rays. The craft is nine metres high and houses forty-eight antennas.

Neutrinos on ice

In October 2014 Katie Mulrey, a postgraduate student in physics at the University of Delaware, travelled to McMurdo Station as part of the NASA-funded ANITA – Antarctic Impulsive Transient Antenna – collaboration to build and launch ANITA III, a scientific balloon that uses the entire continent of Antarctica for neutrino and cosmic ray detection. In the first of a series of posts she wrote for *Scientific American*'s expeditions blog, Mulrey describes the science behind the project.

October 22, 2014 – Right now, somewhere in the Milky Way, a proton is travelling though supernova remnant shock waves, bouncing around in extreme magnetic fields and gaining energy and speed. Eventually the proton will gain enough energy to escape the remnant and travel through the galaxy. It might even head to Earth. If we are lucky it will crash into our atmosphere, creating a shower of millions of particles, just waiting to be detected by a large balloon that will soon be launched over Antarctica.

Information about high-energy cosmic rays and neutrinos sheds light on what are now mysterious processes in the universe. What natural phenomenon can produce such energetic particles? Where are they geographically located? How many high-energy particles can we see? This information will lead to the next generation of high-energy astroparticle physics.

Neutrinos and cosmic rays travel through the universe at incredible speeds. Neutrinos are fundamental particles in nature, and part of the Standard Model of particle physics. The name neutrino means "neutral little one". As this suggests, neutrinos have no charge, and so they are hard to detect.

Cosmic rays are charged particles, commonly protons or other atomic nuclei, that are very energetic and travel at speeds close to the speed of light. They are the most energetic particles ever observed (far greater in energy than anything that can be produced in particle accelerators) and they should produce nearly as energetic neutrinos by interacting with the light left over from the Big Bang – the cosmic microwave background (CMB). We are talking about energies in the 10^{19} electron-volt range and higher. It is almost impossible to grasp what energies this high really mean. The visible light we see tends to be on the order of one electron volt. A 10^{19} electron-volt proton has the same energy as a baseball thrown at sixty miles per hour.

These particles have to undergo extreme acceleration to achieve these energies. Sources of lower energy neutrinos and cosmic rays are fairly well known. Our sun, for example, produces many lower energy neutrinos via nuclear fusion. We don't think too much about neutrinos on a daily basis, but in fact, sixty-five billion neutrinos pass through your fingernail every second. Since low-energy cosmic rays and neutrinos are more abundant, scientists have a better grasp of their sources and composition.

High-energy neutrinos and cosmic rays pose a larger problem. The origin of high-energy neutrinos and cosmic rays is unknown, and more information about these particles will shed light on the most energetic processes in the universe: supernovae, active galactic nuclei, gamma ray bursts, etc. The problem is that they are much rarer than their lower energy counterparts and so very difficult to detect. In order to detect these particles you either need a telescope with a huge detection area, or a huge amount of detection time, or both.

When cosmic rays enter the atmosphere they interact with particles in the atmosphere, creating particle cascades, aka "air showers". A similar phenomenon happens when high-energy neutrinos pass through a dense material such as ice. The cascade development generates electromagnetic radiation that can be seen in many wavelengths, including the optical and radio spectrums.

Detectors tuned for specific wavelengths can be used to detect cosmic ray air showers with radiation from different parts of the electromagnetic spectrum. Enter ANITA, an instrument consisting of forty-eight antennas arranged on a balloon payload to scan the entire continent of Antarctica for radio signals that come from neutrinos interacting in the ice, as well as cosmic rays interacting in the atmosphere and reflecting off the ice. Large detection areas are very hard to build, so we use what nature gave us – a huge sheet of ice, perfect radio detection.

There have been two ANITA flights so far. Both have detected cosmic rays and have been important for neutrino research. This all leads to ANITA III, which was assembled for the first time last summer at a NASA balloon facility in Palestine, Texas (where the climate was very different from ANITA's final destination). Scientists worked for two months to build and test this experiment, which has been years in the making. In order to proceed, the instrument had to pass two tests. The first test was to make sure the payload wouldn't fall apart under its own weight, nearly 2.5 tons. The second was to ensure that scientists on the ground could communicate with the instrument while it was flying around Antarctica. The instrument passed both crucial tests in late August.

So what happens next? We take it all apart: the thirty-foot-high gondola structure that houses forty-eight antennas, the communication equipment, loads of cabling, and the "instrument box", the heart of ANITA, which processes and records the data. Everything is put in huge shipping containers and sent south to Antarctica, where we will meet it in late October when the austral summer begins.

What do we expect to see? Seeing any neutrinos would definitely be awesome. Two things are happening in the high-energy neutrino world: experiments are becoming more sensitive and models are getting better. ANITA's goal is to bridge the gap between experiment and theory. Models put a limit on the number of neutrinos we can expect to see. If we don't see any within that limit it means the models may need to be changed. In that sense, any news will be good news. If we see neutrinos, we can confirm and tweak the models. If we don't, we can put constraints on the models. Cosmic rays, however, are expected to be bountiful.

When Mulrey arrived at McMurdo Station some members of the ANITA team had already started work.

November 11, 2014 – It has officially been two weeks since we've seen the sun set. It's been surprisingly easy to get into a routine here. We work seven days a week to make sure ANITA will be ready on time, so each day feels the same. Every morning we wake up and catch a bus to the CSBF Long Duration Balloon (LDB) facility. It is located in the shadow of Mount Erebus, the most southern active volcano on Earth. It's always smoking to remind us it is alive. We've found out it is much colder out there than it is in town. Thank goodness for all of our cold-weather gear. McMurdo Station is located on land but the LDB is built on the Ross Ice Shelf. On the ride in we drive over the boundary between the two. ...

We are stationed in an airplane hangar. It was empty when we showed up, but day after day it filled up with more equipment as shipping containers arrived.

Now it's time to start building. This is one of the most exciting phases of the experiment. The gondola – the structural frame of the payload – starts out as thousands of pieces of carbon fibre tubing and metal connectors. It is built from the top down. The communication equipment and solar panels are at the very top: the GPS and communication antennas need to be on the top of the payload so they have direct access to satellites. Next come the antennas. They are organised radially into sixteen "phi sectors" and are tilted ten degrees down towards the continent. This way we can see over the ice in all directions.

ANITA consists of forty-eight broadband antennas (180–1200 MHz) and one low frequency antenna (30–80 MHz). The broadband antennas each have two feeds, one horizontal and one vertical. The vertical polarisation is great for detecting neutrinos, since the radio signal propagating through the Antarctic ice has to be vertically polarised to escape without totally internally reflecting. Cosmic-ray radio emission is primarily caused by charged particles in the air shower moving horizontally in the Earth's magnetic field, so the horizontal polarisation is great for detecting them. The signals coming from all the antennas are cabled to the ANITA "instrument box".

The signals we see are very faint. If someone on the ice happens to start a car engine, the signal can reach the payload and look very similar to a neutrino signal, so a lot happens between the time the signal reaches the antenna and the time the data is recorded in the instrument box. The RF (radio frequency) signal first goes through an amplifier, then through a long cable, through filters, splitters, more amplifiers, more cables, and then into our triggering and digitising system. All these components have differing characteristics (times 97 channels!) and need to be carefully calibrated before we can say what our data really means. We do that by sending known pulses into the system and recording what comes out the other side. Our data analysis depends on the exact timing and amplitude of the signals, so calibration is critical. ...

November 23, 2014 – It is important for ANITA to see as much of the Antarctic ice as possible. Neutrinos interact in the ice, so viewing as much land area as possible increases the detection area and makes it more likely we will see them. This means we have to fly pretty high. Flying electronics at an altitude of 115,000 feet presents some unique

ANITA III is transported to the launch site for the sixth and successful attempt at launching it on a huge vehicle nicknamed "The Boss".

problems, however. You would think overheating wouldn't be a problem in Antarctica, but we fly in the stratosphere which has very low air density. Fans cool most electronics at sea level but this technique breaks down when there is very little air. We have to make sure no individual components are generating too much heat or, if they are, that they efficiently dissipate it. All metal parts are connected together with very messy heat sink compound, which facilitates cooling through passive energy transfer. We also take steps to make sure sunlight reflects off the instrument by covering all metal that would conduct heat with reflective tape. The balloon's gondola is painted white to help it stay cool, and all the cables are sheathed in white.

Speaking of the gondola, it is almost complete. We added two more levels of antennas and put all the scientific instruments on the deck of the gondola. Now we have to wear hard hats and harnesses when we need to get into the instrument to fix something. We still need to add two deployable structures to the payload. Size is the limiting factor here. When fully deployed, ANITA is larger than the hangar in which it's being built. This means that some pieces have to be tucked up inside the payload and released once the balloon is in the air. That is always exciting, because if something goes wrong with the deployables during launch there is nothing that can be done about it. In our case we are deploying solar panels and a huge low-frequency antenna. If the solar panels fail we will have only about six hours of battery life.

Another large undertaking this week was the set-up of our remote antenna pulsers. It can be difficult to know what signals we are seeing while we fly unless we send some known signals up to the balloon. We can't do this from McMurdo because there is too much background radiation from wi-fi, walkie-talkies, etc. Two of our team members headed off to Siple Dome, 507 nautical miles from McMurdo Station, to set up an antenna that will send known signals to ANITA from the remote site. Siple Dome is a field camp. Only two people stay there during the Antarctic season. They are responsible for cooking, managing the station, keeping the runway clear and everything else that has to happen. The camp is primarily used as a fuelling station for planes heading further out. ...

While undertaking fieldwork such as this, it is important we keep our knowledge of basic physics intact, so we made sure to spend some time practising the conservation of angular momentum with a bit of back-spinning fun on the ice.

December 15, 2014 – It's another beautiful day in Antarctica and the time has come to launch ANITA. Finding the right date is tricky. Many factors have to fall into place. In order to detect neutrinos and cosmic rays we want to fly over the Eastern Ice Sheet. We detect these particles via their radio emission, and the smooth ice provides a great surface for the radio waves to refract in a predictable way, as opposed to the mountain ranges in central Antarctica and the bumpy coastline.

A polar vortex, forming a giant ring of stratospheric winds, sets up around the South Pole in early December. The air currents in the vortex dictate our flight path. If we launch too early and the vortex hasn't set up properly, we might end up flying over the ocean, which we don't want. We want to get as much flight time as possible, to maximise the exposure time we have for detection.

Each day we check the vortex maps and forecasts that NASA prepares. Once the vortex is up and going, we wait for the weather conditions to be right. The winds have to be low, under ten knots, and heading in the correct direction. Launching a balloon takes up to ten hours so the weather has to be good for a long time. Conditions change rapidly in Antarctica. If we think ten a.m. might be a good launch time the next day, we wake up and prepare at two a.m. We launch pibals to gauge the wind conditions.

ANITA has been six years in the making. Its success depends on the next few days. All the antennas are connected for the last time. Once NASA declares us "flight ready" we don't touch the experiment, so there is no way to be positive we have everything hooked up correctly. This leads to checking, double-checking and triple-checking. The last bits of the payload are painted white to ensure they don't get too hot during the flight.

The balloon is launched with ANITA III attached at the end.

Now we wait for the weather to cooperate. So far we have had five attempted launches. Riggers from NASA's Columbia Scientific Balloon Facility bring ANITA to the launch pad and prepare to unroll the balloon. There is no going back once the balloon is taken out of its box, so that is often the point when the CSBF weatherman makes the crucial decision about whether or not to launch.

We were very close to launching on December 14 but high winds picked up at the last minute. We even had a special visitor come to watch the potential launch: a few hours before flight an emperor penguin wandered out to our facility. It is still a little early to see penguins here so it was a great surprise. ...

January 5, 2015 – Launch day did finally arrive [December 17, 2014]. The winds were low, and the forecasts were promising for a great neutrino-detecting balloon launch. This was ANITA's sixth attempt at launching. Everyone was more than ready to bid the balloon farewell. Until the balloon and payload are airborne the scientific team has little to do on launch day. The launch is in the hands of the CSBF rigging, electronics and weather teams.

The balloon is launched from a huge vehicle named "The Boss"; behind it everything necessary for flight is laid out in a long line. First you have The Boss holding the payload on a crane. Next comes a parachute that will be used to bring the payload back down to the ground when the flight is over. Attached to the parachute is the balloon... . The balloon is made out of a material similar to a plastic grocery bag, so it is very delicate. Once it is taken out of the wrapper there is no going back, because trying to repack the balloon might cause some damage. We were waiting for the news that the wrapper had been opened, then we knew launch was imminent.

Only the tip of the balloon is filled with helium. That is enough to lift the two-ton ANITA to thirty-six kilometres, where it will take data. The entire balloon will inflate to the size of a football field once the balloon rises to lower density atmosphere. Helium is inserted into the balloon through two long plastic tubes that get tied off after the correct amount is added. That balloon can lift the whole payload so a lot of work goes into keeping it on the ground before the launch.

At launch time The Boss starts driving the balloon and payload in the direction of the wind. It lets go of the payload and the balloon begins to rise in the air. The timing is very hard to get right: if you hold the payload too long the balloon will pull too hard and could tear; too early and you drop the payload on to the snow. Talk about exciting! There was a lot of cheering as we watched the balloon float away.

Now it's data time. There are hard drives on ANITA that are constantly writing data but we can't access that until the flight is over, and then only if the payload can be recovered. (In some cases the payload will fall into water or be covered with snow so we can't get it.) We record lots of radio signal data, only a small fraction of which is really a neutrino or cosmic ray. We send down data we think might be really important via satellite. At the beginning of the flight we also have "line of sight" access to the balloon, meaning we can talk directly to the instrument with radio communication before the balloon goes over the horizon. After that we rely solely on satellite communication. ANITA collaborators watch the data stream 24/7 to see if everything is working properly. We monitor the temperature of the instrument, voltage and current levels, and plots of the radio data coming in.

If the signals from our forty-eight antennas match up in the right way, we know we have a neutrino. We spend months after the flight carefully analysing the data to be sure we know what we saw. ANITA has almost made one circle around Antarctica. If we are lucky, we might get two more circuits. The more data we collect, the better.

For so long, we thought of ANITA as a mammoth science instrument looming in front of us. It looked so small floating away into the atmosphere. It's amazing that that payload will shed light on the highest energy particles on Earth.

ANITA III was in the air for twenty-two days and nine hours, during which it completed one and a half circuits around Antarctica before falling to the ground. In early January 2015, a team from Australia's Davis Station located and retrieved the structure's instruments. As this book went to press the data was still being analysed.

From "Neutrinos on Ice" by Katie Mulrey: scientificamerican.com, 2014/2015.

Global Barometer

In 2002 a section of the Larsen B Ice Shelf the size of Rhode Island collapsed. This huge body of ice, on the east side of the Antarctic Peninsula, had been stable and intact for more than 10,000 years. A recent NASA study suggests the rest of the ice shelf will collapse by the end of the current decade.

The Antarctic Peninsula is one of the fastest warming places on the planet: average temperatures have increased by 2.5°C over the past fifty years. Not only there but all around the Antarctic coast ice shelves and glaciers are showing signs of instability. As the warming ocean destabilises coastal ice, the massive ice sheets and glaciers that cover the interior of the continent are flowing more quickly towards the coast, melting at an average rate of 120 gigatonnes of ice each year. Some scientists now believe the entire West Antarctic ice sheet is headed for inevitable collapse, with an associated global average sea level rise of up to 3.6 metres.

The atmospheric CO_2 that's warming our atmosphere and oceans is affecting ocean chemistry too. Ocean acidification, often called 'the other CO_2 problem', is creating increasingly inhospitable conditions for many Antarctic marine species. At the same time the warmer temperatures – in sea and on land – are inviting to new species, such as the king crabs that are marching towards Antarctica along the ocean floor, or the grass seeds and insects arriving as stowaways with unwitting Antarctic scientists or tourists.

Antarctica has become a global barometer for change, and much of the research there is now focused on understanding the extent of this change and what it might mean for the future. Sediment and ice-core drilling projects are helping scientists learn more about past climates and the changing extents of the major Antarctic ice sheets over time as predictors of possible futures for us in a warming world. Others are studying what growing ocean acidification will mean for different species. Others still are collecting baseline data – documenting marine and terrestrial ecosystems now, so we can see what effect the changing climate, and invasive species, have over time.

But there are still more mysteries arising that have yet to be solved. How fast have ice shelves and ice sheets collapsed in the past, and how will any changes in them contribute to future sea-level rise? How will the recovering ozone hole affect atmospheric circulation and climate? And, in this warming climate, why does the extent of Antarctic winter sea ice continue to grow?

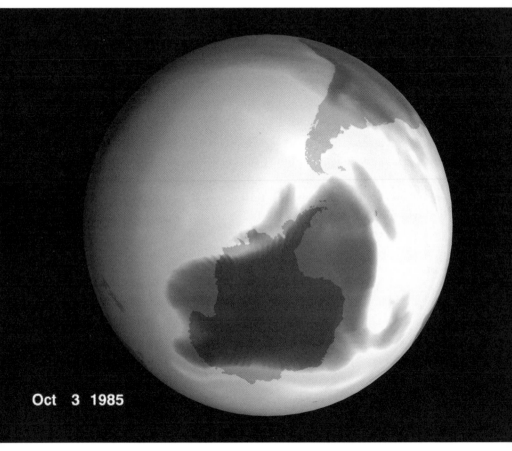

Oct 3 1985

NASA's image of the ozone hole above Antarctica, October 1985. As the ozone layer thins, more ultraviolet light reaches Earth's surface, increasing the risk of sunburn, skin cancer and eye cataracts.

Earth sans sunscreen

In 1985 the science magazine *Nature* published a letter by
three scientists from the British Antarctic Survey in which
they showed that a pattern had emerged in the ozone levels
above Antarctica: each spring the levels had started to fall
dramatically. Jonathan Shanklin, Joseph Farman and Brian
Gardiner had made the discovery by analysing atmospheric
data recorded at the British Halley Research Station on the
Brunt Ice Shelf in East Antarctica. In 2009 Jonathan Shanklin
reflected on the surprising find.

The formation of the Antarctic ozone hole is a graphic demonstration
of how rapidly we can change the atmosphere of our planet. There
are many other environmental issues facing us today and we must link
them together to understand and debate the underlying causes, rather
than treat each issue in isolation. Antarctica is a wonderful continent.
Glaciers carve their way to the sea, where the waters teem with penguins
and whales. Although seventy percent of the world's fresh water resides
in the polar ice cap, the continent is a veritable desert, with liquid
water in short supply. The frozen ice takes on many shades, from the
brilliant white of freshly fallen snow to the deep indigo at the bottom
of a gaping crevasse. This land of contrasts is where the Antarctic
ozone hole was discovered.

Ozone is a form of oxygen similar to the gas that we breathe, but
with three atoms instead of two. This makes it highly reactive, and in
high concentration it is a toxic gas. When formed by air pollution near
the surface it can trigger asthma attacks, but high in the atmosphere
it forms a protective sun-shield. This is the ozone layer, a region
from about ten to thirty-five kilometres in altitude, where the natural
concentration of ozone is highest.

Ozone forms above this level in the stratosphere through the action of ultraviolet sunlight on oxygen gas, and in the process the most harmful ultraviolet radiation is totally absorbed. Some ultraviolet light does reach the surface and the intensity is controlled by the amount of ozone – the more ozone the less ultraviolet, and vice versa. With a thinning ozone layer more ultraviolet light reaches the surface, exposing us to greater risk of sunburn, skin cancers and cataracts of the eye.

Ozone observation in the Antarctic began over fifty years ago with the International Geophysical Year of 1957–58. As part of this scientific endeavour a network of observatories was set up across Antarctica, several of which measured ozone. One of the first to report was the British research station Halley, and the results from the first year of operation showed a surprising difference to those from the equivalent latitude in the Arctic. This was soon recognised as being due to a different stratospheric circulation in the atmosphere above the two poles: in the north the circulation is relatively complex, while in the south it is relatively simple, with a strong long-lasting winter polar vortex or a large-scale persistent cyclone.

Ozone observations at Halley continued using the same type of instrument, the Dobson ozone spectrophotometer. Designed in the 1920s by an Oxford professor of physics, Gordon Dobson, it remains the standard for ozone observations today. The instrument uses ultraviolet light from the sun coming through the ozone layer to measure the amount of ozone. It is very much a manual instrument, and the calculations required to extract the ozone amount from the observations are quite complex, to the extent that in the 1970s a stack of unreduced observations began to build up.

When I joined the British Antarctic Survey, one of my first jobs was to write computer programmes that would process the observations once they were entered into electronic form. Making sure that the entered data was correct was the first part of the process, followed by verifying the software.

At about the same time, concern was growing that spray cans and the Concorde supersonic airplane could destroy the ozone layer. When the British Antarctic Survey held its open day, it seemed a good opportunity to reassure the public that the ozone layer above Antarctica had not changed. Surprisingly, however, the data seemed to show that the springtime ozone values of that year were much lower than they had been a decade earlier. I had yet to process the intervening data, but once this was done it was obvious there was a systematic effect, giving rise to the paper that Joe Farman, Brian Gardiner and I wrote, announcing an unexpected effect over Antarctica.

Elsewhere in Antarctica, other observatories had continued to make ozone measurements on a sporadic basis, but they lacked the long-term continuity of the same instrumental technique that was available at Halley. This was a key factor in our discovery, and set a valuable lesson for monitoring the environment. In addition, the centre of the ozone hole is often offset towards the Atlantic, allowing Halley to start making observations several weeks before the sun rose high enough at the South Pole.

Once the paper was published in *Nature*, satellite data was reprocessed to reveal an "ozone hole" over the southern continent. While satellites give an excellent overview of the changes within the ozone layer, ground-based observations are still needed to provide them with an accurate calibration.

Today we know that this ozone hole is caused by chlorine and bromine from ozone-depleting chemicals such as chlorofluorocarbons (CFCs) and halons. The reason for the particularly severe ozone depletion over Antarctica lies with its stable polar vortex, which makes the Antarctic ozone layer roughly ten degrees colder than that in the Arctic. This means that unusual clouds form widely in the Antarctic ozone layer during the winter, and chemistry on the surfaces within these clouds conditions the ozone-depleting chemicals. When sunlight returns, very efficient photocatalytic reactions that destroy ozone take place.

The Montreal Protocol has been a very effective response to the shocking and rapid change in the ozone layer over Antarctica. Now ratified by all but one of the United Nations member states, it is having a clear effect in reducing the amount of ozone-destroying substances in the atmosphere.[3] CFCs and allied substances are, however, very stable, so their atmospheric concentration drops very slowly and will not reduce to pre-ozone hole values until at least 2070. It is likely to be several more years before we can be confident that the ozone hole is shrinking and many decades before springtime ozone levels return to those of the early 1970s. One unintended consequence of the reduction in ozone-destroying substances has been its significant effect on reducing global warming as the substances are often also powerful greenhouse gases.

Treating the ozone hole was relatively straightforward, with both general acceptance of the need to change and the possibility of alternative products. Another environmental symptom – that of climate change – is currently generating much debate, but the amount of greenhouse gases in the atmosphere is rising at the worst-case rate predicted by the Intergovernmental Panel on Climate Change (IPCC). In addition, there are many other global symptoms of environmental stress, ranging from water and food shortages and fishery collapses to deforestation habitat destruction, among others.

When a doctor treats a patient with an illness, it is essential that all his or her symptoms are taken into account in making a diagnosis. It must be exactly the same when we are looking after the health of our planet. My diagnosis is that we must urgently debate and act on reducing our effect on the planet, otherwise ever more symptoms will appear.

---

3 Universal ratification was achieved in September 2009.

Such reduction could be achieved through decreasing the consumption of our planet's resources, particularly reducing consumption among the developed nations; but we are also likely to need to reduce our own numbers if we are to sustain a healthy planet in the long term. How to do so is the big debate we must urgently conduct if we are to avoid a fate such as the inhabitants of Easter Island, who used up all their resources. Unfortunately, these warnings, like those of Cassandra, are unlikely to be heeded and it may require a major disaster before action is taken. The United Nations is one forum where the debate should begin.

From "Unlayering of the Ozone: An Earth Sans Sunscreen" by Jonathan Shanklin: unchronicle.un.org, August 2009.

The Clean Air Sector Laboratory at Halley Research Station where Rhian Salmon and her colleague Stéphane Bauguitte studied Antarctica's atmospheric chemistry for sixteen months, including through the polar winter.

Waiting for the polar sunrise

Rhian Salmon worked as an atmospheric chemist at Halley Research Station, one of the most remote bases in Antarctica. On her first visit over the summer of 2002–03 she helped oversee the construction and commissioning of a Clean Air Sector Laboratory, or CASLab, from three prefabricated shipping containers. She returned the following year, this time travelling by ship from the UK with about thirty different boxed-up machines designed to measure various molecules in the air. She and her colleague Stéphane Bauguitte stayed at Halley for the next sixteen months, running these machines and capturing data about Antarctic atmospheric chemistry. One of the main reasons they had to winter over was to be there (and ready) for polar sunrise, when new physical and chemical reactions are kicked off after 105 days without sunlight.

In 2004 blogging was still quite a new phenomenon and the internet had not yet reached Halley. Finite amounts of data could be sent and received only at set times of the day. Salmon had a data limit of 1MB per month. She used this to send emails to her brother, who published them on his blog, and to write letters to her grandmother. The letters would be emailed to British Antarctic Survey headquarters in Cambridge and posted to her grandmother's address. In return, her grandmother's handwritten weekly letters would be posted to the headquarters and faxed through to the wintering radio officer, who passed them on to Salmon.

This is an adaptation of one of Salmon's blog posts from early June 2004, not yet halfway through the long polar night. The posts were often written to an imagined but familiar "you" but when she returned to the UK she discovered her readership was far larger than she had ever imagined.

While writing my weekly letter to Granny yesterday, I found myself stumped for words. Stumped for material more like. What to say? Nothing has changed since last week; there is no news to report. It's still cold, it's still dark, I still go to the lab to work and I still live with the same seventeen people I have for the last four months, and in the same physical environment where I've been for seven months. (There were sixty people on base until February but since then no one has been able to arrive or leave, nor will they be able to until November.) We have had no visitors, no mail, no dramatic new events unfolding. Don't get me wrong, we're happy here, laugh a lot, party, play games, watch videos, read books, have a million bizarre conversations and a few serious ones too. But there's not much to report.

I feel as though I've already told you anything worth telling but that you somehow expect more. You want drama and excitement, courageous battles with the weather, brilliant diversions from near-miss incidents, comrades going mad with winter depression, and skin shrivelling up through lack of fresh fruit. But it's not like that. When you're here it just makes sense. But maybe that means I'm taking it all for granted, so this morning I thought I'd pay a bit more attention to my daily routine.

The sunshine lamp – an attempt at preventing Seasonal Affective Disorder – floods my room with squintingly bright light around seven a.m. thanks to an old fridge defrost-timer that the wintering electrician rewired for me. Unimpressed, I crawl back under my duvet and hide my head. At about 7.45 my alarm clock goes off, jolting me back out of dreamland for just long enough to hit the red button on the top. I am parched. At 8:45 I finally pull myself out of bed, down the glass of water I left there the night before, climb off the top bunk and into my issued clothing: thick thermal socks, green moleskin trousers and a thermal top. Yum. Oh god, you're thinking, is she going to go through the *entire* day in this kind of detail? No. Well, maybe. We'll see.

The point is, it's really hard to get out of bed in the morning (and also the driest place on Earth). I've never found it easy. I know past friends and housemates are already laughing at this, but it's really, *really* hard. One guy here even asked if gravity was stronger at the poles due to some effect of the rotation of the Earth (which, as an aside, is not such a dumb question, and one we have debated extensively at dinnertime).

There is definitely a winter lethargy about the place. It's not that we don't get stuff done. We just get it done at our own pace, in our own time. There are plenty of theories about why this is. The most widely accepted is that daylight acts as a reset switch in our body clocks, helping us to remain in approximate twenty-hour synchronicity with each other. It's well documented that Antarctic winterers can rapidly lose synchronicity from each other – ultimately resulting, in extreme situations, in social breakdown. Which is precisely why I insist on dragging myself out to work around nine a.m. and try to finish before dinner – something I have never done at home when I have had the freedom to work whatever hours I choose. Ironically, I have never lived such a disciplined life as here, where the only social constructs are those respected by the wintering contingent.

I digress. I get up, have breakfast, chat over a cup of tea, peg out, tog up, and go out to the lab:

Get up but don't have a shower, as water (from snow, which we manually dig into a melt tank) is limited and I like to shower after I've hauled myself indoors, sweaty and tired, at the end of the day.

Breakfast is toast or cereal. Bread varies according to the person on night shift, apples and oranges are still around but the quality is a lottery, and I often make yoghurt, so that might go in a bowl as well.

Tea is obvious. *Chat* is usually related to articles in the newspaper that arrives overnight via the radio officer, who prints it out for us; it has a scarily biased perspective on everything. I am now an expert on reality TV's comings and goings, David Beckham's sex life, and where the royal princes will be holidaying next.

Peg out – well, there's a peg-out board where you put your name next to the place you're going to, and write it in a book, along with the time you expect to be back. The daily "gash" person (we rotate that role) keeps an eye on the book throughout the day and tries to find you if you're out beyond your self-provided curfew. It's a little big brotherish but could save your life.

Togging up – now I could write a whole book on Antarctic clothing but in short here's what I add to my clothing in the boot room: mukluks on my feet (big moon boots that are steel toe-capped and almost knee-high), a thin fleece top, all-in-one orange padded overalls, a balaclava, a neck warmer, a dead-rabbit mad-bomber hat, a thick "windy" cotton smock that blocks out the wind like no other material I have ever experienced, goggles, thin glove liners, and large mitts. Plus an emergency backpack with more clothing if I'm going to the CASLab, and a radio worn like a vest under the outer layer to keep the batteries warm.

Okay, so I'm togged up and ready to leave the building. It's probably nine-thirty. As I am leaving I realise I need to go via the Simpson platform to pick up some solutions I had mixed on Friday, and some more deionised water for various machines at the lab. (We call these outer buildings "platforms" because they're raised about two metres off the ground, and are jacked higher each year due to an annual accumulation of about two metres of snow.) Last summer a wet chemistry facility was built on the Simpson platform: three of our instruments use liquid methods, but for reasons of space, safety and practicality there is nowhere to prepare liquids at the CASLab where my machines are. In order to ensure we don't just measure our own pollution, the CASLab was built about 1.3 kilometres upwind from the Laws platform, where we live, so if you go via the Simpson the journey is a bit longer. In bad weather or reduced visibility we have to walk via the Simpson anyway as this is the way the hand lines are routed.

Leaving the Laws I remember that I bust the binding on my skis last week so will have to walk. Since I'll be taking lots of liquids with me, I take the orange pulk sledge and chest harness, dump the emergency bag inside the sledge, and pull the lot to the Simpson platform. It's not a long stop. I pick up five litres of dilute sulphuric acid in a glass bottle, three litres of sulphanilamide solution, a few pots of pre-weighed powder, three litres of deionised water, fifty millilitres of acetone, and some crunchy granola bars. I'm sure I've forgotten something but it'll have to wait.

It's cold and windy outside, bordering on a storm, so before leaving I make sure there are no bits of skin peeking through my headgear. I got a reasonable-sized patch of frost nip last week on my left cheek and eye. It's now healing nicely and I don't want to re-expose it. I load everything into the sledge and start the slog to the lab.

But what's twenty-hour-hour darkness like, you're asking. Well, at this time of day it's dark still, always dark. Today was quite cloudy so there wasn't even the joy of stars to carry me along the commute. The moon is waning but still shedding quite a lot of light and the lights from the lab are bright enough that I can follow them without needing a head torch. Another brief storm last week dumped a lot of soft snow unevenly in lumps and bumps across the ice so the walk is quite hard going. I fall over a couple of times and struggle when the sledge catches on sastrugi – the hard, bumpy but beautiful surface of the snow. I'm pulling less than fifteen kilograms on the sledge but it feels like a lot more when we're going in opposite directions.

What's the point of all this? The big picture? Well, it all goes back to climate change. For a long time, ice cores have been used to try and gain an understanding about past climates. If we know about past climates, we might be better able to predict future climates – or at the very least appreciate to what extent recent dramatic changes in the climate are due to human influence, or merely part of a cycle that has ancient timescales. So it's worth taking a moment to think about how those ice cores form. Snow falls on the ground. It is light and fluffy and full of air. As more snow falls, this earlier snow settles and becomes more compacted; with time the weight of snow above becomes so heavy the early snow starts turning into ice. Any air mixed in among the snow becomes trapped in bubbles in the ice. Centuries later these bubbles become even more compacted and the ice is totally transparent, ancient, a memory – like fossils and lake sediments – of days when dinosaurs walked the planet.

Unlike dinosaur fossils, the ice has a timescale: depth. It can be used to understand not only what ancient climates were like but also how the climate has changed with time. The ice is not just water, nor is trapped air just nitrogen, oxygen, carbon dioxide and argon: there are other chemicals and particles that may suggest the presence of forests or deserts. More recently formed ice shows a record of lead appearing from petrol emissions, with that signal reducing again when unleaded fuel was introduced.

Around the time of the industrial revolution, methane and carbon dioxide concentrations appear to soar. And so, it seems, does the temperature of the planet. Obviously temperature can't be measured directly from the ice … but the ratio of different isotopes of oxygen trapped in air correlate directly with temperature. It's all proxy data and correlations, you see, theories and assumptions, but all this information does seem to be in quite startling agreement. And that is the basis of a sound scientific theory.

So why am I here? An atmospheric chemist. (I don't know the first thing about ice and past climates, although I'm learning.) Well, the

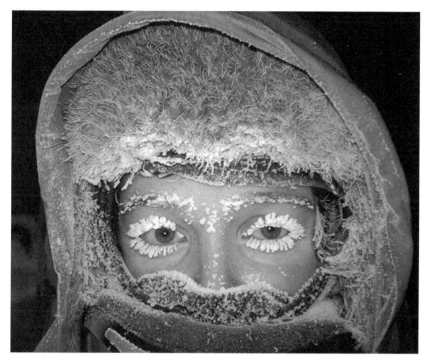

Rhian Salmon in balaclava, neck warmer and "dead rabbit mad-bomber" hat.

above story makes the massive assumption that, after snow falls to the ground, the chemical composition of the air mixed within it doesn't change. Remember, the snow is light and fluffy precisely because it is full of air. The assumption is that this air just gets pushed deeper and deeper, unchanged in chemical composition, until it eventually becomes trapped in bubbles in ice. Then thousands of years later some random scientist drills a very deep hole, takes a slice of ice, and analyses the air that is trapped within it ...

The "air not changing" argument is fairly sound once the air is trapped in deep cold ice. At that stage it's hard for those pockets of air to go anywhere (physical) or do anything (chemical). However, while they're still inside snow there's no reason they can't flow around the snowflakes, diffusing upwards – back into the open atmosphere – or downwards, towards the ice. And while this happens, the pockets of air that move within the snow might pick up molecules in the snow and free them into the open atmosphere, or react with molecules in the open atmosphere and deliver them to the snow. In addition, the surface of snowflakes inside the snow might be a wonderful opportunity for chemical reactions to occur that wouldn't ordinarily happen in the open atmosphere.

It's all physics. And chemistry. And maths. The point is that to truly understand the research results that tell us about the composition of air trapped deep, deep down in the ice, we need to understand what has happened to that air between being in the free atmosphere and being trapped in bubbles several centuries later. So rather than just needing experts in ice cores and past climates, there is also a need for people who study present-day air, and the processes by which air is transferred from the atmosphere to the snow, then to bubbles in ice, which is why the British Antarctic Survey has employed two atmospheric chemists to spend a winter here. And why I'm sucking air out of tubes buried vertically in the snow (to investigate what's happening twenty metres below the snow surface), and sampling melted snow (to investigate processes just below the snow surface), and flying blimps (to investigate air twenty metres above the snow surface), and why we're bouncing light between the CASLab and a mirror four kilometres away (to investigate air just above the surface).

It seems to take an age to get to the lab. In reality it is probably about twenty or twenty-five minutes but I'm sweating loads when I finally arrive at ten-thirty and am relieved to get indoors. There are outdoor checks I usually do upon arrival but today I'm too tired and will cool off rapidly, so I decide to do them when there's a bit more light outside.

Thankfully, everything seems fine once I get unwrapped and my breath back. First I check the telescope room, but it's too cloudy to see any return signal from the mirror bouncing light back at us from four kilometres away. On a clear day this is one of the most satisfying experiments as the return beam looks like a star on the horizon and is easy to check if it's nicely focused on to the optical fibre at the end of the telescope. We're looking for absorbance of light by a range of atmospheric molecules and their relatives, mainly halogen oxides and the nitrate radical. We don't expect to see much chemical activity during the winter months but it's important to measure this so we know how it compares to the increased signals when the sun returns. Year-round studies of this nature in Antarctica are still pretty special.

The next machine – or "instrument" to use the proper word – measures formaldehyde in the air. It seems to be ticking away nicely but one of the reservoirs is running low, hence the sulphuric acid I hauled with me. That should last another week or two now, although the other liquid it uses is also running out and I'll have to make a new batch this week, maybe tomorrow afternoon. The HONO instrument is next. (We call instruments according to what they measure. In this case the liquid's chemical formula, HONO, is easier to say than its real name, nitrous acid.) I refill the helium bag and feed the empty reservoir next to it with the three litres of deionised water.

Next to this instrument is a gas chromatograph that measures peroxy acetyl nitrate, or PAN, and I notice its water reservoir, used for circulating coolant, is also low. This instrument stands in a rack. On the bench next to it is a gas chromatograph that monitors non-methane hydrocarbons – we give this the snappy name "NMHC GC" – that is working well at the moment. It needs a blank run-though so I set up a nitrogen cylinder and change the programme accordingly.

The last of my instruments to come online is around the corner next to the formaldehyde monitor; this measures peroxides and will hopefully be running soon. While checking these five machines I have barely had to move my feet at all as they are crammed together around one human-sized patch of floor.

Sometimes this routine feels like watering plants – going from machine to machine, checking if each is too wet or too dry, in need of more liquids or gases, or maybe the occasional prune or extra dose of attention. At other times, on good days, I feel like the conductor of an orchestra, enjoying all these different machines working in parallel, none particularly interesting on its own but collectively providing an incredible understanding of all the amazing processes happening in the air. On bad days, though, it's more like being a kindergarten teacher with a roomful of noisy complaining children, each seeking more individual attention than I have time to give (and each of which would, I am sure, be happy and helpful if only I had time to attend to their individual needs).

I make a note of the various solutions I need to prepare next and spend the rest of the morning paying the machines a bit more attention – running calibrations and blanks, troubleshooting things I don't understand, and sending emails back to the UK if I need more help.

The truth is that there's nothing very exciting happening right now in terms of the data we're collecting. It's very cold and very dark and this means that our molecules are pretty dormant – there's not much energy in the system to stimulate new chemical or physical processes. On the whole we are either measuring consistently low and stable concentrations (as expected) or nothing at all, which isn't to say these molecules aren't around, but if they are they're in concentrations lower than our instruments can measure. In fact, a few years ago there were no instruments that could have measured Antarctic concentrations at any time of year – the air here is just really, really, really clean, which means really, really, really low concentrations of the things we're looking for. One of the reasons this project is exciting is that many of

Rhian Salmon and meteorologist Craig Nicholls up a mast four kilometres from the CASLab, assembling an array of quartz retro reflectors. The reflectors act as a perfect mirror for bouncing back a light beam sent from the lab.

these molecules have never been measured before in Antarctica, and we actually have no idea if we will be able to measure them. The simple (or not so simple) act of trying to measure them, whether or not we succeed, will be a result in itself.

Right now, in the middle of winter, we're not seeing much and don't expect to see much. But all that will change (we hope) when the sun comes out … and really, that's why I'm here: to catch the sunrise. Or rather, to catch the explosion of chemistry we think happens when the sun reappears. That's right – sixteen months on an isolated Antarctic base for one week of data. So I'd better make sure all our machines are working when that day comes.

Sunrise is possibly the most exciting day of the year in Antarctica. Everything changes, and that includes molecules in the air. This is especially true for a coastal location like where we are – it's no coincidence this year-long experiment is being carried out here. During the winter, Antarctica roughly doubles in size due to the freezing of sea ice. In spring that sea ice begins to melt, a process that involves the formation of pools of really salty water. These pools contain high concentrations of bromine and iodine which, when zapped by sunlight, release a series of highly reactive species into the air; these are bigger than atoms but smaller than molecules, the intermediary things that are formed during a chemical reaction. They subsequently react with molecules in both the air and the snow around and create new molecules that weren't around in winter … which we will hopefully measure in real time, catch on camera as it were. And *that* in a nutshell is why I'm here.

Sometimes it's good to be reminded of the big picture.

It was a nice morning all in all: not too hectic and I made some good progress. The lab is cramped and noisy though, so I was hoping to make it back in time for lunch at one p.m. Around midday I venture outside for the outdoor work, hoping it might have warmed up a bit. I go on to the platform and collect a small plastic jar for sampling snow. It's dark today so I need a torch to find the pole where we sample; it's about 100 metres from the lab. Once back on the platform I check the pressure in various gas cylinders, look at the air inlets on the roof to check they're not blocked, and pick up some freezer blocks (stored outside) to keep solutions indoors cool.

By the time I get back inside after less than ten minutes, the end of my nose and tips of my fingers have gone white and are screamingly cold. The pain doesn't get better as I warm them up with the palm of my hand. Damn – I must have not covered up properly, I've been nipped again. It's time to leave. I want to leave. I'm hungry and not looking forward to the journey back. I get dressed again, this time with another balaclava and even bigger mitts from the emergency bag, grab the bag, empty bottles, and a twenty-litre jerrycan full of waste chemicals that need disposing back on base; it will eventually be shipped back to the UK.

The air is dark, I can't see any definition in the snow, and the sledge is heavy and difficult to pull on this lumpy surface. I'm being obstinate: I should really dump the sledge and pick it up tomorrow but some days you just don't want to let the weather beat you, so I keep going. Within a few minutes my goggles are steamed up and frozen. I had put them on in a different order this time that obviously doesn't work: I can't see squat. Without them I can see the lights of the Laws platform (home) but risk getting more frostnip.

I realise I'm not in a good mood, and that it's unfortunate today of all days is the one I decided to document in writing. Some days, when it's clear, the walk home is a joy – I chat to the stars, dance with the snow, sing to myself. But today it's a slog. It's cold, it's dark, it's one o'clock for crying out loud. I can't see a thing and am having to navigate using a tiny slot between my furry hat, balaclava and wind-jacket hood. I try wearing the goggles and hanging on to the hand line but that is even more ludicrous. I know I'll be okay: I have no doubt about my safety or ability to get home. I just wish I could be there now.

Forty minutes and a couple of stumbles later I crawl indoors and collapse for a couple of minutes. Drama queen. The gash dude did call me on the radio just after my last stumble. It was kind of comforting knowing that if I hadn't replied they would have come out looking for me. Nice to know the system works.

A bowl of tomato soup and a bacon roll later I'm laughing about it. I've checked the weather data and it's not as bad as it felt: only −31°C with fifteen knot winds. It felt a lot colder: with wind chill it translates to −42°C. Still, at this time of year it usually has to reach −40°C or wind speeds above thirty knots to get much of an ooh and an aah around the table. What a wuss!

I've warmed up but my nose is sore, my cheek raw, and I don't fancy returning to the lab. All is fine there and I have plenty I can be getting on with on the Simpson platform for the afternoon. But then I realise I might have left the gas store open when my fingers got frostnipped, with the intention of closing it later. Damn. It would probably be okay for the night, but if the wind picked up the bottles would get cold and the store fill with snow. It's my colleague's day off (we alternate weekends and Mondays) but, bless him, he offers to go out and check for me in the afternoon. I am so grateful. None of my face fancies more of that journey again today.

Truth be told, I'm a bit bored with this stormy weather. It's not big enough to be exciting but too big for any good outside activity – the way you get bored with grey days and dull rain in Britain. Last week was much nicer. I was on melt-tank duty and often sat on the snow mound after the tank was full, looking at the night sky in the morning or enjoying the red glow if it was an afternoon dig. I even went for a walk on Saturday, the weather was so calm.

I spend the afternoon replying to work-related emails, fixing a pipette and analytical balance in the wet chem lab, and preparing some more chemicals. That winter lethargy seems to have crept back in again.

Dinner's at six, after that we have our usual banter and then I pick up a new book. My last were *Love in the Time of Cholera* followed by *Touching the Void* and I am now starting *Oranges are not the Only Fruit*.

At eight we have our Monday night double bill of the TV series *24*, and then, after my heart rate calms down, I plug in my laptop and start writing this. It was a fairly typical day I guess, if a bit colder than usual. Some days the weather is better but the lab is much worse. I'd rather have the former by far. I thought there was nothing to report, but I guess all this feels normal only if you do it every day.

The Antarctic field component of the project on which Rhian Salmon was employed – CHABLIS, studying the "CHemistry of the Antarctic Boundary Layer and the Interface with Snow" – ran from December 2003 to February 2005. It resulted in thirteen peer-reviewed publications, including ten articles in a special issue of *Atmospheric Chemistry and Physics* and a paper in *Science* that announced the highest concentrations of iodine oxide measured anywhere in the atmosphere. The CASLab has since been moved to a new research site and continues to host year-round atmospheric chemistry experiments for the British Antarctic Survey.

Adapted from "An Average Day" by Rhian Salmon, originally published on felixsalmon.com in 2004 and now at smilingfootprints.com.

Antarctic sponges and anemones and a sponge-eating starfish on the sea floor of McMurdo Sound.

Life under ice

"Beyond the familiar images of a frozen wilderness there are luxuriant ecological communities on the sea floor," marine ecologists Paul Dayton and Simon Thrush write in their introduction to this description of life beneath the sea ice of McMurdo Sound. In contrast to the stark white landscape of the continent, the sea floor is a vibrant collage of colourful invertebrates and silvery fish, with some of the highest densities of animals anywhere in the world.

A first-time diver in McMurdo Sound slides anxiously through a small hole in the sea ice, pauses to accommodate to the low light, and is immediately struck by the remarkable water clarity. During much of the year underwater visibility can exceed 200 metres, which changes the distance perspective for divers used to murkier conditions.

The diver looks down and the sea floor looms close: around Ross Island it is volcanic rock and gravel. Benthic life is conspicuously depth-zoned. The first zone is relatively barren except for starfish, giant nemertean worms and occasional fields of purple solitary hydroids. Descending further, the diver encounters increasing numbers of sea anemones, starfish, sea urchins, soft corals, solitary and compound ascidians, and scattered sponges.

At around thirty metres the bottom is covered with sponges of many shapes and colours, and some are astonishingly large. If the diver descends to about fifty metres she may find the sponges giving way to large aggregations of bryozoans – colonial animals similar to corals. Above, the ceiling of solid ice looks like a stormy afternoon sky with patches of light and dark, the pattern being determined by surface snow cover and the accumulation of algae under the sea ice.

Once the diver sees her escape hole she relaxes and looks around to see fish perched on their fins and perhaps a seal gliding overhead and peering down with large curious eyes.

As she ascends, she ponders the differences between this habitat and those in temperate regions: the lack of seaweeds, the way fish hug the bottom and lunge from one spot to another, and the depth-related change from relatively bare sea floor to areas covered with sea anemones, to sponges, and then to bryozoans. She pauses beneath the ice to admire the clusters of large ice crystals, or platelets, often covered with brown diatoms and small crustaceans, and the small silvery fish swimming through the crystals.

Before she finally ascends through the hole, the diver looks around and sees small translucent comb jellies and purple sea butterflies fluttering by, while in the distance large colourful jellyfish pulsate along, dragging long tentacles. She looks down through the clear water one last time to the blue bottom and its interesting animals. She is hooked; for the rest of her life she will dream of returning. ...

McMurdo Sound is one of the most fascinating coastal marine habitats in diving depths on Earth. It has an incredibly wide range of densities and diversity of benthic animals. In some parts of Ross Island, on the eastern side of the sound, the densities of sea-floor organisms – up to 155,000 individuals per square metre – are some of the greatest recorded anywhere in the world. In contrast, on the other side of McMurdo Sound communities of sea-floor animals are more akin to those of the bathyal deep sea, with up to 10,000 individuals per square metre.

One reason for the big difference is ocean productivity. The currents along the west coast of Ross Island often sweep down from the north, laden with nutrients and high in primary production derived from coastal open-water areas, thus providing abundant food for animals. Meanwhile, the western side of McMurdo Sound is bathed by waters that have been under the massive Ross Ice Shelf for a very long time, have had little primary production, and are nutrient-poor.

One exception to this is the Antarctic scallop, *Adamussium colbecki*, which is abundant along the western coast of the Ross Sea. In productive waters around Terra Nova Bay, it reaches densities of more than fifty per square metre in water thirty to fifty metres deep. Further south in McMurdo Sound, at places such as New Harbour and Salmon Bay, it is also found at these densities, but only in the shallows near the ice walls that form where the sea ice meets the beach.

The presence of these anomalously large and abundant animals in such a nutrient-poor area is a result of freshwater run-off from glaciers and snow banks that melt during summer. When the freshwater enters the sea it freezes in the cold salty sea water, creating an ice wall that prevents further freshwater reaching the sea. In the relatively warm freshwater that is trapped behind the ice wall phytoplankton begin to grow, turning the water green and so helping it to absorb heat from the sun. This high-productivity water flows over the ice wall at the edge of the land and into the sea beneath the ice. In this zone along the ice wall, scallops reach their highest densities. They become rare in water deeper than thirty metres. ...

Adaptation to an extreme environment usually produces quirks of nature that provide insight into how far organisms must go to survive in such places. For example, the low temperatures and high salinity of polar waters make it hard for organisms to use calcium carbonate to construct shells or other skeletal structures. Thus, most of the bivalves found in these ecosystems are tiny, and even the larger ones such as the Antarctic scallop have very thin and easily crushed shells. The small fishes seen resting on the sea floor (*Trematomus bernacchii*) and on the undersurface of the sea ice (*Pagothenia borchgrevinki*) are living in water only a couple of degrees warmer than a domestic freezer. The body fluids of these fish are saltier than the surrounding sea water, providing some antifreeze protection, but the fish have also evolved complex proteins and sugars that further lower their freezing point. These two factors limit ice formation in their tissues.

Low temperatures produce low metabolic rates in most animals and plants, which is an advantage in an environment that is dark for months on end and in which food is only plentiful for a few months each year. A low metabolic rate means organisms can tick over using very little energy, but when food is available they are able to rapidly take advantage of it. In other words, despite the very cold water the overall growth rate of sea-floor animals is limited by food rather than temperature. This creates a strange situation in which many ecological processes seem very slow in comparison to those in temperate New Zealand, but any food fall, or a slightly warmer summer, produces bursts of biological activity, the effects of which can linger for many years.

Quick change in slow systems is illustrated by sponges in McMurdo Sound. Long-term studies have shown that benthic populations, at least at some sites, are affected by inter-decadal or El Niño oceanographic processes that influence the formation of ice. There are multi-year cycles of the emergence of supercooled water from under the ice shelf. This very cold water freezes around animals growing on the sea floor and eventually becomes sufficiently buoyant to rip them from the floor. In some years "anchor ice" completely strips the bottom of its fauna in depths down to thirty metres. Then there are periods of almost no anchor ice formation, and at such times in the intermediate depth zone there may be a massive recruitment by fast-growing *Homaxinella* sponges. This in turn contributes to the success of the sponge-eating starfish, which may heavily impact upon the sponge community for decades.

Sponges are a spectacular component of the sea-floor communities of the Ross Sea. There are about 300 species and most are glass sponges, which have silica spicules to provide structural support. These sponges add structural complexity to sea-floor habitats and provide refuges for many species. Even when sponges die, their spicules remain as a mat on the sea floor. The resulting open but complex matrix radically increases the sea floor's surface area and provides an ideal habitat for many animals, diatoms and other microscopic plants.

Sponges are suspension feeders, although the discovery of macroscopic algae in the tissue of a variety of species suggests that algae help sponges meet their energy requirements when food resources are scarce by providing them with carbohydrates. An even more unusual and fascinating physical phenomenon has been uncovered in the hexactinellid sponge *Rossella racovitzae*. This species has spicules that act as natural fibre optic cables, conducting light to the diatoms within the sponge's body. Apparently the sponge benefits when the diatoms die and decompose. ...

Despite the cold and disturbance by ice, the coastal sea-floor communities of the Ross Sea are diverse and in places highly abundant. In many ways the ice has helped us to understand these complex ecosystems by creating gradients and patterns in distribution and abundance. There is much to learn about these ecosystems and Antarctica offers an opportunity to integrate our understanding across many sciences and large scales of space and time. Given the growing evidence of climate change, this information becomes increasingly important. The cold and the low-density human habitation mean many historical features remain intact and provide the opportunity to step back in time. Cores of ice from the glaciers and the ice shelves, and cores of sediments from the sea floor, have revealed the region's complex history of climate change. Learning about nature is an adventure, especially in Antarctica. The knowledge gained in this adventure is vital as the pressures on the continent and southern oceans grow, but it also provides a view of nature from a relatively pristine environment.

From "Chapter 32: Antarctica" by Paul Dayton and Simon Thrush, in *The Living Reef: The Ecology of New Zealand's Rocky Reefs* edited by Neil Andrew and Malcolm Francis: Craig Potton Publishing, Nelson, 2003.

Scientific research vessel the *Joides Resolution*, which is capable of
recovering marine sediment cores from more than 2,000 metres below
the ocean floor, revealing information about Earth's past.

Thirty-six million years in Antarctica

The East Antarctic Ice Sheet is the world's biggest and oldest ice cap. It formed around thirty-four million years ago as Earth transitioned from a warm greenhouse world, with no or minimal ice caps, to a colder world with substantial ice caps, lower sea levels, and glaciers.

Sediments on the sea floor record the climate and the ocean conditions at the time they were deposited. By drilling deep off the coast of East Antarctica, collecting sediment cores and studying them, scientists can learn more about the changing extent of the East Antarctic Ice Sheet over time, along with other data such as the temperature and the levels of carbon dioxide in the atmosphere and ocean – information that helps them build models to predict future climate changes, ice melt and sea level rise.

The 2010 Wilkes Land Expedition focused on learning about conditions in the area over the last thirty-six million years. Marine geochemist Rob Dunbar of Stanford University reported from the scientific drilling ship *Joides Resolution*.

February 1, 2010
Temperature 0°C, wind 35 kts, 5-metre swells
At Site U1356, Hole U1356A
Position: 63°18.6139'S, 135°59.9397'E
Water depth: 4003 metres
Core depth (penetration into the seabed): 1004 metres
Total weight of over 3 miles of pipe hanging from the ship: >650,000 pounds!
Aboard the *Joides Resolution* off the coast of Wilkes Land, Antarctica.

Wow, what a week! We just finished retrieving our final core from the bottom of a drill hole more than one kilometre in length. We've now recovered and described sediments that range in age from a few million to more than thirty-six million years old, all in the span of about nine days.

The first sediments that came up told us what Antarctica was like when the ice sheet was as it is today. Then we saw evidence of a much warmer time, and then a colder time before that – a time when flotillas of icebergs carried rocks and debris from the Antarctic continent out over our drill site, dropping this debris as they slowly melted. Even further back in time, more than thirty million years ago, we began to find evidence of much warmer waters – and for the first time no evidence of large ice sheets.

We also began to see sediments that may have come from Tasmania or other parts of Australia. Even though we are now thousands of kilometres away from Australia, back in time, thirty million years ago, Tasmania and Antarctica were much closer, perhaps only hundreds of kilometres apart. Plate tectonics since that time have carried Australia to the north, while Antarctica has remained more or less anchored at the South Pole. So not only do our sediment cores tell us tales of past warm climates (and perhaps give us hints as to what lies ahead in our greenhouse future), they also tell us new things about the science of plate tectonics.

The weather here changes fast. Yesterday, we had a warm (well, maybe 4°C) and sunny day – our first sunshine in over three weeks. After shift, everyone went outside to feel the warmth of the sun and to see blue skies and blue water.

I was up for shift at midnight. It was calm and cloudy with snow at four a.m. but by nine a.m. it was blowing thirty-five knots and we now have waves over twenty feet high. The change in weather happened while we were retrieving our very last core.

Now we will spend another two days here doing something called "logging". Logging is when we send instruments down to the very bottom of the drill hole after we remove the metal pipe that supports it. These instruments measure the properties of the rocks we drilled through. By doing this we can piece together the sections of sediment we actually saw and described, even across breaks in our core that may have been caused by problems with the drilling. Altogether, we usually recover about fifty percent of the rocks we drill through. Some of the softer or stony units just can't be cored and recovered very easily so this kind of work – logging – is very important for us. ...

It's great fun out here and the time is going by very quickly, just about as fast as the short Antarctic summer.

February 6, 2010
Temperature 3°C, wind 5 kts, 1-metre swells
Position: 66°33'39"S, 136°59'E
Water depth: 1000 metres
Exact location: The Antarctic Circle

Yes, we crossed the Antarctic Circle today. It is perhaps only the third time this ship has ever done so. All points south of the Antarctic Circle experience at least one day a year of total darkness, and likewise one day a year when the sun never completely sets. We are now in early February so the sun does set but only for four hours and it never gets really dark. As a member of the night shift out here, I love this ... I get up at eleven p.m., come on shift at midnight. The sun sets around one a.m. and rises again around five. I get to see both, and when the weather is good the colours are spectacular.

We are now working at one of our shallow continental shelf sites, U1358. We just finished the major site for which I am the lead scientist. This site was cored very successfully. The water is 1000 metres deep and the spot we cored is like a big dish at the sea floor, with lots of small sediment particles drifting into it.

The sediments accumulate at the rate of two centimetres every year and leave an annual layer – a summer deposit made up of microscopic plants and a winter layer made up of dust and silts carried by the wind and the ice. We can see each layer and each layer represents one year. It looks as though we can count these layers back over 10,000 years. The record may not be perfectly continuous, we don't know yet, but we do know we have 470 metres of layered mud to work on and that it will tell how the sea ice and temperature of Antarctic surface and deep waters has changed on a year-to-year basis for many thousands of years.

Everyone on board worked long hours to get this site completed, many for eighteen to twenty hours each day, so when we have a transit day to another site we get to rest, but we also have a chance to cross the Antarctic Circle. Everyone is excited and a bit relaxed, both at the same time. The weather is sunny and warm today but tonight we expect a big storm to begin, with winds gusting to over sixty knots and waves as high as twenty-five feet. It might last two to three days, a problem for us as it is difficult to work in such stormy conditions. I'll let you know how it turns out.

February 16, 2010
Temperature –2°C, wind 30 kts, 3-metre swells
Transiting back to Site U1359
Position: 64°34'S, 140°30'E
Water depth: 3700 metres
The scene outside: 2 days of storms and lots of icebergs

Our latest drilling target is in an area where sediments that document the transition of Antarctica from the "hothouse" to the "icehouse" can be easily reached at shallow depth beneath the sea floor. We drilled for eighteen hours and then had to pull the drill pipe up out of the hole and reposition the ship to avoid a large iceberg that was heading straight for us. When the iceberg had passed the weather started to deteriorate. Our forecast was for sixty-knot winds and big seas so we headed north out of "iceberg city" to ride out the storm in deep water away from icebergs and sea ice. The forecast was true to its word – we had waves up to thirty feet and winds over sixty knots for more than twenty-four hours. But we had great iceberg viewing on the way to our WOW – Waiting On Weather – point so I'll write something about them and how they fit in with our project.

The Antarctic ice sheet is always accumulating new snow that gradually turns to ice. For the ice sheet to remain the same size it must either melt or release ice to the ocean as icebergs. In parts of Antarctica some of the ice is in fact melting, but most of the ice loss that maintains the continent at its present state occurs through the calving of icebergs. Most icebergs calve off ice tongues and ice shelves – areas of concentrated ice flow at the coast.

Imagine that the ice is draining off the high parts of the continent by flowing down small ice drainages to form mighty rivers – but rivers of ice in this case. These vast rivers move slowly, only a few tens to hundreds of metres each year. When they reach the coast, the ice flows out into the ocean, where it begins to float wherever the water is deep enough. In some cases, this is where the water is over 500 metres deep and the ice is over 560 metres thick. Floating ice shelves or ice tongues are influenced by winds and ocean currents. They begin to melt if the water is warm enough but they mostly break up to form icebergs.

Many of the icebergs here off Wilkes Land came from the Ross Ice Shelf – the world's largest ice shelf. It is over 1,500 kilometres away in the Ross Sea, but icebergs travel great distances in the Southern Ocean.

The water is cold and they drift with the ocean currents, for decades in some cases. As they drift they melt a bit below the waterline and become rounded. Sometimes they flip over and this rounded part is then visible. Icebergs often collide and gouge away at each other, or they list over at an angle and slowly fall apart. This means that icebergs come in all shapes, sizes and textures.

The ice at the base of the ice sheet often carries sediments: boulders, gravels, cobbles, and sands. When these parts break off and begin to float they form "dirty" bergs, with dark rocky layers intermixed with the clear blue ice. The debris that falls from these dirty bergs accumulates in sediments at the seabed. When we see gravels or sands in otherwise fine-grained sediments, we know this debris was transported out over the ocean by ice. In fact, the presence or absence of ice-rafted debris is something we keep close track of in the cores we are collecting on this trip. This tells us whether Antarctica was generating lots of icebergs and therefore had at least some kind of ice sheet in the past. Conversely, when we see sediments that do not contain this debris we know we are looking at a record from a time when Antarctica was much warmer.

February 26, 2010
Temperature –3°C, wind 5 kts, 1-metre swells
Drilling at Site U1361
Position: 64°24.6'S, 143°53.2'E
Water depth: 3470 metres
11 more days!

Hi folks! We've been busy the past nine days. We are coming to the end of our work window here in Antarctica. Summer has ended, the nights are much longer, and next Tuesday we must begin our transit back to Hobart, Australia.

Twenty- to thirty-million-year-old marine deposits, drilled from sediments 750 metres below the sea floor, in water 4,000 metres deep; the stones fell out of melting icebergs and dropped to the sea floor. The scale is in centimetres.

We made our last attempt to get into one of our continental-shelf drilling sites but were stopped once more by sea ice. All the ice here is frozen sea water, as opposed to glacial ice.

We finished drilling deep into the seabed at Site U1359, encountering sediments and rocks that span much of the last thirteen million years. For several hundred metres we found layers of rock that alternated between green and brown over distances of two to four metres. The green parts of the cycle showed evidence of intense mixing of the sediments by animals living at the sea floor, while the brown parts showed something quite different – very well-developed laminations that could not have survived if the sediments were mixed. There were different materials in the layers too. Some contain more shells of diatoms, the main plant in the surface waters of the ocean. We don't yet know what these cycles of green and brown represent but we think they reflect the continued glacial–interglacial cycling of Earth's climate many millions of years ago.

Today we are in a very cold period in Earth's long-term history, and have been for most of the past several million years. Because Earth is already quite cold, when we have glacial–interglacial cycles we see large ice sheets coming and going at both poles – in Antarctica, Greenland, and over large parts of North America and Scandinavia. This waxing and waning of polar ice is driven by small changes in the shape of Earth's orbit around the sun (it changes from an ellipse to more circular and back again over about 120,000 years), the tilt of Earth's axis (it wobbles a bit over 40,000 years), and the exact seasonal timing of when Earth is at its closest approach to the sun. All of these "orbital" changes impact how much sunlight reaches Earth, as well as when and where it warms it seasonally. Sometimes, Earth is in an orbital configuration that produces warm winters and cool summers – a combination that usually allows ice sheets to form and grow. Some tens of thousands of years later the opposite occurs – warm summers and cool winters – which can cause ice sheets to rapidly melt. In today's cold world, these small changes have big effects as the polar regions are cold enough to allow large ice sheets to form and last through the warmer periods.

Antarctica has been covered in a large ice sheet for many millions of years because of this overall cooling. The ice sheet still waxes and wanes along its margins, but it is always present in the continent's interior. However, before about two and a half million years ago there was no permanent ice sheet in the north polar regions – it was simply too warm. Further back in time, the Antarctic ice sheet was much smaller than it is today, but it was still dancing to the rhythms set by Earth's orbit.

What we were seeing at our last drill site, and what we are looking for at our latest site, is evidence of how these glacial–interglacial cycles affected the Southern Ocean, and how they in turn may have been different because the planet was a little bit warmer than today. By studying this we can learn more about how small changes in the planet's temperature can affect things such as ice volume, sea ice extent, and the productivity of the ocean. This is directly relevant to understanding our greenhouse future. Although all the climate variability that occurred millions of years ago was "natural" – in other words, not caused by people – the strength of the signal that caused these past changes (the orbital changes in this case) is not very different from the strength of the signal we expect from the man-induced increase in carbon dioxide levels in the atmosphere.

The poles are a great place to study both natural and man-induced changes in Earth's climate because of a phenomenon called polar amplification. We know from many hundreds of studies of the past fifty million years of climate change that, whenever Earth warms up, the poles warm up more than the planet's average. The converse is true for times of cooling. We don't yet fully understand why but based on these studies we shouldn't be surprised that the poles are warming up very quickly today, at a rate greater than in the tropics or the temperate belts. The cores we collect on this trip have the potential to tell us more about when and why polar amplification occurs.

I'll send one more blog from this trip once we have cleared our last hole and are heading for Hobart. Sixty days is a long time to be at sea and a long time to be working every day for fourteen hours or more. We are all excited to get home.

March 5, 2010
Temperature 12°C, wind 10 kts, 1-metre swells
2 more days at sea...
Aboard *Joides Resolution* in transit to Hobart, Tasmania.

The work of the ship ended as quickly as it had started nearly two months ago. We finished drilling Site 1361 and logged the hole. The drillers tripped 3500 metres of pipe and prepped it for storage as the ship will not drill again until July, off the coast of British Columbia. Everyone on board is absolutely brain-dead from the non-stop grind of twelve-hour shifts day after day, but happy as well. We've completed most of our objectives and made some exciting discoveries. When we did not meet with complete success it was always because of weather and ice, either encroaching sea ice or fields of icebergs so thick we had no chance to pass.

Now we have some days in transit. These days are filled with meetings to design our post-cruise research. We will all spend much more time at home working on the cores than our actual days at sea on this expedition. Some of the methods we will employ are expensive and difficult and we have recovered nearly 2000 metres of core. This means we must carefully select the intervals we will study so we can answer the most important questions about Antarctic climate change as quickly as we can. For some of us, the analytical work will extend over the next four years. Then other scientists will work on these cores for decades to come. They will be stored in a vast library of ocean cores in College Station, Texas, the IODP core repository, where they will be available to scientists from all over the world.

What I like most about these days in transit is going off shift. I no longer set my alarm to wake at eleven p.m. The two shifts mingle at meals and in the labs, almost strangers at first as we have not seen much of each other for more than seven weeks.

Working groups between the shifts assemble to design research strategies and timetables. I will lead a group that will make oxygen isotopic measurements of the small shells of amoeba-like organisms called foraminifera. Forams, as we call them, live for about four weeks during the brief Antarctic summer. They build their tiny shells out of calcium carbonate, the main mineral that makes up limestone. By measuring the ratios of two types of oxygen in the carbonate we can tell the temperature of the water in which the forams grew.

We will make these analyses on forams that were living in Antarctic surface waters hundreds, thousands, and even millions of years ago to see how warm the water was next to the Wilkes Land coast. We already know from our microscope work on board that this part of Antarctica has been very warm at times, maybe ten to fifteen degrees centigrade warmer when we go back thirty-five million years. The foram work will help tell us exactly how warm the waters may have been during more recent periods when we know that the ice sheet became much smaller. The results will help us predict the behavior of Antarctic ice in the future.

What a trip it's been. ... As we pull ever closer to Hobart we are very much aware that we are simply reaching the end of the beginning.

Adapted from a series of blog posts by Rob Dunbar at *Ice Stories: Dispatches from Polar Scientists* on icestories.exploratorium.edu, 2010.

Antarctic sea butterfly *Limacina antarctica*. The delicate shells of
this important species are at risk of dissolving as the ocean becomes
increasingly acidic.

Hermaphrodite butterflies
and acid seas

American biologist James McClintock's passion for Antarctic
marine biology began in the 1980s, inspired by ocean voyages
to the Kerguelen Islands in the southern Indian Ocean and
to the Antarctic Peninsula, and summers spent at the United
States' Palmer and McMurdo Stations. McClintock has a special
interest in echinoderms – starfish, sea urchins, brittle stars,
feather stars and sea cucumbers – and has led research projects
into ocean acidification, marine invertebrate reproduction and
chemical ecology. Here he writes about ocean acidification –
the "other CO_2 problem" – and the impact it could have on
the marine ecosystem.

The most common sea butterfly in the seas of Antarctica – the lovely
Limacina antarctica – is poised for catastrophe. The same Southern
Ocean that has for aeons provided both habitat and nourishment is
turning acidic and threatening its survival. Its shell, thin as a human
hair and measuring half an inch across, is sculpted of elegant whorls
of translucent mineralised aragonite. Ocean acidification – which can
damage or dissolve these thin fragile shells – is the enemy.

What these swimming snails give up in body size they make up for
in sheer numbers. A single bathtub of Antarctic sea water contains
up to 5,000 individuals. A cubic mile of the surface of the Southern
Ocean houses twenty-seven trillion butterflies. Their mind-boggling
abundance renders this mid-sized planktonic organism (mesoplankter)
a classic keystone species. The production and decay of their shells
is of sufficient magnitude to play into the global carbon cycle – a
combination of biological, chemical and geological processes involving

the exchange of carbon between all of the world's natural systems, including soils, rocks, lakes, oceans and atmosphere. In the Southern Ocean the construction of countless shells of sea butterflies requires the uptake of carbon-rich aragonite, one of two common crystal forms of calcium carbonate. Upon the snail's death, the shells slowly sink, transferring the carbon within them to the deep sea floor. Once they reach the bottom the shells either dissolve or form fine silt sediments called pteropod ooze, that ultimately become rock and clay.

Shelled sea butterflies such as *Limacina antarctica* have unique reproduction and feeding habits. When they are young, the butterflies are all males. As they age and approach sexual maturity they change into females, an intriguing transition known as protandric hermaphrodism. But if this were not odd enough, just before changing males mate with males, each storing away the mutually exchanged sperm until it has completed the final transition to becoming a female.

As females, the snails then use the stored sperm to fertilise their eggs and release the resultant embryos into the sea, where they develop through two phases of swimming larvae. First-stage larvae are shaped like spinning tops, bristling with tiny hairs they use for swimming. Second-stage larvae are tiny versions of juveniles-to-be, complete with shells but equipped with a thin flap of tissue covered with tiny hairs, called a velum, that they use for swimming and capturing food particles. When a larva prepares to metamorphose into a juvenile, it absorbs its velum, providing nutrients and energy for this final stage of development.

As juveniles, shelled sea butterflies launch into a lifetime of herky-jerky swimming generated by rapid flaps of wings. After several months, fully grown and with mature reproductive organs, they join the rank of adults. Surviving a year or more, Antarctic sea butterflies are long-lived compared to their temperate and tropical cousins.

Their feeding strategy is as odd as their sex life. Secreting a mucous web, they suspend themselves in the water column like parachutists, the vast expanse of the webs dwarfing their individual body size. Shaped like a sphere, the webs serve to provide buoyancy and capture prey. Like fishers, the shelled butterflies periodically reel in their sticky webs, consuming any prey entangled in the mucous.

Shelled sea butterflies are quick to abandon their elaborate webs when disturbed. As such, vast numbers of deserted webs drift in the plankton and serve an important ecological role as breeding grounds for bacteria and as food for smaller planktonic organisms. With shelled sea butterflies as numerous as stars in a galaxy, even their faeces become important in the web of life. The sinking faecal pellets play a key role in transporting organic material from the surface of the Southern Ocean to its deepest depths.

Victoria Fabry, a biological oceanographer, is an expert on ocean acidification and its impacts on shelled sea butterflies. In 2007 Vicky and several of her students travelled to McMurdo Station to study what happens to the shells of Antarctic sea butterflies when they are exposed to conditions of elevated ocean acidification. The team placed the shells of sea butterflies into sea-water tanks adjusted to the acidity and mineral levels predicted to occur in Antarctic sea water by the year 2100. In less than a week the central coils of the delicate translucent shells turned opaque and they began to dissolve. At three weeks the shells were completely opaque, and at one month the once-smooth symmetrical coils of the shell had become crooked and ragged. After six weeks, the investigators ended their experiment – the once intact shells were but ghostly shadows of their former selves, no longer suitable domiciles for sea butterflies. Over a span equivalent to a tenth of their life expectancy, the shells of the sea butterflies were no more.

In a second dramatic illustration of the vulnerability of polar sea butterflies to ocean acidification, scientists used a powerful atomic microscope to observe cracks and valleys in the micro-architecture of the outer surfaces of shells of a subarctic sea butterfly. Remarkably, these living individuals had been placed in water that mimicked conditions of future acidified sea water for just forty-eight hours. Yet sea butterflies face not hours but a lifetime in an acidified sea. Polar shelled sea butterflies are in peril.

In addition to their shells transporting carbon to the sea floor, the sea butterflies play a critical role in food webs, their flesh providing essential nourishment for animals further up the food chain, including a vast array of jellyfish, fish, marine birds and baleen whales. Should sea butterflies become unable to produce or maintain their fragile shells in a future ocean, the prospects for the myriad sea creatures that exploit this bountiful buffet are bleak.

Scientists have called ocean acidification "the other CO_2 problem"... [and it] has yet to receive the same level of attention that newspapers, magazines, television and the internet have paid to the inescapable role of carbon dioxide as a potent greenhouse gas – the original CO_2 problem. But its back-burner status is changing. Marine scientists are voicing their collective concerns about the potential impacts of ocean acidification on natural ecosystems and fisheries, and popular science writers are taking note. ...

The chemistry behind ocean acidification can be simply explained. When sea water absorbs atmospheric carbon dioxide in the atmosphere it immediately reacts with water, releasing hydrogen ions that lower the pH of the sea water. The pH scale provides a measure of the concentration of hydrogen ions and thus the level of acidity. The lower the pH value the higher the acidity. Since the beginning of the second industrial revolution, the average pH of the world's oceans has been reduced from pH 8.2 to pH 8.1. This decrease does not sound like much, but a decline of one-tenth of a pH unit in human blood can cause acidemia, a condition that can lead to coma and death.

This similar small decline in the pH of sea water actually represents a thirty percent increase in the acidity of the world's oceans. Scientists predict that if humans continue to burn fossil fuels at the current rate, by the end of the twenty-first century the world's oceans will have an average pH of 7.8, making them a stunning 150 percent more acidic than at the onset of the second industrial revolution. Unfortunately, the amount of time required to undo the damage of ocean acidification is considerably greater than the time necessary to cause it. Even if by the end of the century humankind were to cease expelling carbon dioxide into the atmosphere, it would take tens of thousands of years for the pH of the world's oceans to return to pre-industrial-revolution levels.

A second important outcome of atmospheric carbon dioxide reacting with water is that a significant portion of the hydrogen ions released combine with dissolved carbonate ions in sea water to form bicarbonate ions, thus depleting the global pool of oceanic carbonate ions, the building materials used by a myriad of sea creatures to manufacture their calcium carbonate skeletons or shells. Thus the world's oceans, saturated with the carbonates aragonite and calcite, are now predicted to become limited in carbonates over the next fifty to one hundred years.

The dwindling carbonates will challenge the ability of marine organisms to build and maintain their skeletons and shells. Among calcified marine organisms on the Antarctic Peninsula shelf, those competing for the declining pool of carbonates include coralline red seaweeds, both flattened and erect forms hardened by a skeletal matrix; soft corals, which lack the external skeletal cups of reef-building hard corals but have skeletal needles (spicules) that provide rigidity; clams and snails, which generally have thin outer shells; brachiopods, shelled and clam-like, but with unique filter feeding structures called lophophores; and starfish, brittle stars, crinoids, sea cucumbers, and sea urchins, with skeletons of microscopic or platelike ossicles and spines. ...

A brittle star, an echinoderm, found at Cape Bird in 1971. Since then research has shown that brittle stars and starfish, both with skeletons of magnesium calcite, are the Antarctic marine creatures most vulnerable to oceanic acidification.

As with climate warming, Antarctica is in many respects the Earth's most well-suited natural laboratory in which to study the "first effects" of global ocean acidification, due to two factors. First, Antarctica has naturally low concentrations of carbonate ions – attributable both to carbon dioxide dissolving more easily in ice-cold sea water and to unique patterns of the Southern Ocean's currents mixing with one another. As a result, lower levels of carbonates such as aragonite and calcite, with which creatures build a skeleton or shell, occur in Antarctic sea water.

Second, the skeletons and shells of Antarctic marine organisms are generally very weakly calcified. I have picked up Antarctic clams and snails and without so much as a gentle squeeze accidentally crushed them in my hand. Some shells are so thin that you can see right through

A sea urchin found under the ice at Cape Evans in 2001. Like brittle
stars, sea urchins are at peril from growing levels of acidification in the
Southern Ocean.

them. This may be in part the result of calcification requiring more
of an energy investment at low temperature, or of the evolutionary
outcome of living in an environment with no crushing predators. Most
likely it is a combination of both factors. The bottom line is that weak-
shelled Antarctic marine invertebrates, living in seas that already lack
the skeletal building blocks they need, will be the first and foremost
to face the consequences of an increasingly acidic sea.

The field of studying Antarctic ocean acidification is so young that
marine scientists are scrambling for even the most basic information
about how vulnerable calcified marine organisms actually are. My
beloved echinoderms – starfish, sea urchins, brittle stars, feather stars,
and sea cucumbers – comprise such a group of marine invertebrates.
Echinoderm skeletons are made of magnesium calcite, a mineral even

more vulnerable to ocean acidification than pure calcite, or, especially, aragonite. The higher the ratio of magnesium to calcite in an echinoderm skeleton, the more vulnerable it is to dissolving. Skeletons of temperate and tropical echinoderms examined to date have levels of magnesium sufficient to classify them as highly vulnerable. Is the same true of Antarctic echinoderms? A review of the scientific literature indicates that the skeletal mineral composition of only a single Antarctic echinoderm is known to science.

How can scientists evaluate the vulnerability of Antarctic echinoderms to ocean acidification if they don't even know their magnesium calcite levels? Jason Cuce, a graduate student working with Bill Baker during the 2009 field season at Palmer Station, helped to make up for some of the dearth of information. Due to some equipment issues, Jason's chemical ecology research project had drawn to a premature completion, which meant Jason was available to join a team of biologists collecting fish for their physiology studies aboard the [research vessel *Laurence M.*] *Gould*. Led by Professor Bill Detrich of Northeastern University in Boston, the biologists caught a variety of fish by either deploying baited traps or towing a large net called a trawl along the sea floor behind the ship. The trawl often surfaced with a healthy variety of starfish, brittle stars and sea cucumbers, and on occasion sea urchins, and feather stars in its mesh. In the spirit of generosity so legendary among Antarctic scientists, Professor Detrich extended a warm welcome to Jason Cuce to join on a series of week-long fishing cruises to collect these echinoderms so the team could find out just how vulnerable their skeletons were to ocean acidification.

Jason returned from these excursions with a wide variety of samples for skeletal analysis, and these were augmented with collections made by divers near Palmer Station. They were shipped frozen to us at the University of Alabama at Birmingham, where they were thawed for skeletal analyses and dissected. A strong solution of bleach disintegrated the various body parts, rendering sparkling clean skeletons that were subsequently shipped to a laboratory in Canada for mineral analysis.

The results came back several weeks later: every single one of the twenty-six species in our study fell into the "high magnesium calcite" category. Antarctic echinoderms were indeed highly vulnerable to ocean acidification. Among the groups of echinoderms examined, the brittle stars and starfish had the highest ratios of magnesium to calcite and were, therefore, the most vulnerable to an acidic sea. This heightened vulnerability is important because brittle stars, and especially starfish, are keystone predators whose habits influence how Antarctic sea floor communities are structured. Removing them from the equation would destabilise the food chain. ...

Agencies around the globe have recognised how important it is to understand the impact of ocean acidification. I was invited to participate in the first meeting between all the key US organisations – the National Science Foundation, the National Oceanic and Atmospheric Association, the US Geological Survey, and so on. American scientists of all kinds – from biological oceanographers concerned about the fate of the world's plankton, to marine ecologists examining the fate of coral reefs, marine chemists and geochemists deciphering the impacts of acidification on the chemical and mineral properties of sea water, shellfish biologists lamenting the decimation of oyster hatcheries, and even environmental sociologists deliberating about how best to educate the public – delivered fact after fact pertaining to the growing risks that acidification poses to our oceans.

From *Lost Antarctica: Adventures in a Disappearing Land* by James McClintock: Palgrave Macmillan, New York, 2012.

This fish *Pagothenia borchgrevinki*, commonly known as a borch, has evolved to endure the sub-zero temperatures of the Southern Ocean.

Fishing in Antarctica

Marine biologist Victoria Metcalf first visited Antarctica in 1999 on an ecotourism voyage. Since then she has been to the ice six more times, either flying to McMurdo Sound to carry out scientific work or as cruise director of ship-based expeditions to the Antarctic Peninsula or Ross Sea. Most seasons she has been based at either New Zealand's Scott Base on Ross Island or Mario Zucchelli Station, the Italian base in nearby Terra Nova Bay.

Antarctic sea ice, typically about two metres thick, filters the non-stop summer daylight but still provides enough light for creatures that live in and beneath it. The undersurface of the ice is not, as you might expect, smooth and flat, but beautifully and intricately scalloped and fluted. During the southern summer it is coated in green and brown algae, food for many other members of the Antarctic food web such as copepods and krill.

Cruising this domain in schools, chasing these small crustaceans or other zooplankton, are lean, ghostly pale silver fish with black smudge markings. These *Pagothenia borchgrevinki*, or "borchs", are a species of Antarctic notothenioid fish. The perch-like suborder dominates the Southern Ocean thanks to an evolutionary adaptation that has given the fish a way to endure the sub-zero temperatures. Beneath the sea ice the water is typically around −1.9°C, a temperature that would be lethal to most species. Notothenioids survive thanks to an antifreeze molecule that is present in high quantities in their blood. This keeps them alive but it also causes problems by making their blood thicker. As a result, notothenioids, and especially the icefish family, have developed a variety of adaptations to their blood system to ensure that oxygen demand can be matched by supply. These adaptations include an increased number of capillaries, widening of the blood vessels, and increased heart size and stroke volume.

The antifreeze in these notothenioid fish was discovered by a marine biologist, Art DeVries, who first visited Antarctica in 1961. DeVries worked out that the fish don't freeze because they have a suite of proteins in their blood; we now know these as antifreeze glycoproteins, or AFGPs. Ice crystals are the borch's enemy. The fish either swallow ice crystals floating in the sea water, or the crystals enter the fish's bloodstream through small lesions in their gills or skin. If an ice crystal finds its way into the fish, the proteins bind to each facet of the crystal, making it difficult for the crystal to grow.

These handy antifreeze proteins are produced in the fish's pancreas and then secreted into its stomach and oesophagus. However, decades after they were discovered, it is still unclear how they find their way into the fish's bloodstream.

Another unfurling story is what happens to the ice crystals once they have become bound to the proteins. Those in the digestive system are probably eliminated in the faeces. However, blood-borne, protein-bound crystals accumulate in the spleen and no one yet knows what happens to them there. A long-standing theory was that summer temperatures eliminated the crystals within the fish. However, in a bizarre twist, the antifreeze molecules have also turned out to be anti-melt – in both lab and in vivo experiments the bound ice crystals remain in a solid state, even when temperatures are increased to above freezing point. AFGPs are not only live-saving, but also life-threatening in that lifelong accumulation can lead to an early death.

The evolution of antifreeze in these polar fish was the molecular innovation that allowed them to not only survive in the Southern Ocean but evolve into many species as waters cooled. When extensive sea ice formed fifteen million years ago and other species became extinct, an opportunity emerged for the notothenioids to spread into, and adapt to, new ecological niches. These early fish were, though, bottom-dwelling, or benthic. To head higher in the water column, fish usually rely on a swim bladder they can fill with air. Since the notothenioids lacked a swim bladder they needed to become buoyant by other means. Species such as the borchs lost scales, replaced some of their cartilage with bone, and deposited large amounts of fat over their body. Other notothenioids, such as toothfish, infused their vertebrae and muscle tissue with fat droplets to provide lift, and reduced the amount of mineralisation in their scales, making them lighter.

Icefish have taken this even further: they have no scales and have replaced most of their skull with cartilage. And that's not the only reason this family is special – they also have no red blood cells and no haemoglobin protein in their blood, making their bodies ghostly pale and translucent.

The focus of my research has been trying to identify how Antarctic fish transport fat around their bodies. How do these fish achieve a sense of balance when they have two competing needs – fat for buoyancy and fat as a primary source of energy? It turns out that all the Antarctic notothenioid fish I've studied lack a blood protein called albumin, which was previously thought to be ubiquitous in animals with backbones. Instead, they use high-density lipoprotein, HDL, to transport fat. Whether other species of Antarctic fish also lack albumin is yet to be determined.

I am also interested in whether their fat metabolism system is different to that of other fish, and whether any differences are tied up with adaptation to the cold. What will happen to these fish as global warming heats their ocean? Their ability to cope may centre on their cell membranes. A key component of thermal adaptation is the ability to modulate the ratio of saturated to unsaturated fatty acids in response to gradual temperature change, enabling both oxygen flow and flexibility to be maintained. To maintain fluidity at low temperatures notothenioids have high amounts of unsaturated fat in these membranes.

The Antarctic notothenioids, while supremely adapted to the cold, live in a stable environment that experiences very little temperature change year-round. One of the questions we are asking is whether these fish, in becoming highly specialised to this environment, have lost the ability to modulate membrane saturation levels as water warms. We are investigating this in both the active borchs and the more sedentary and the less hardy *Trematomus bernachii*, or "bernachs", and comparing their ability to alter membrane saturation to that of some of their non-Antarctic relatives.

Carrying out fieldwork on the ice is intensive: it typically involves four- to eight-week stays in fairly extreme conditions. Most of my work involves daily fishing forays out on to the glistening sea ice by Ski-Doo or Pisten Bully. As a team of one I have to rely on others – including fellow marine biologists and Scott Base staff – to help me fish, but the biggest pool of talent comes from visiting media, distinguished visitors and Antarctica New Zealand board members. Each trip involves drilling or redrilling fishing holes using a motorised Jiffy drill. In the past we have sometimes had to drill through as much as five metres of ice. Using a Jiffy drill, with its cantankerous engine, to push through each metre of ice is not easy: with cold, fumbling fingers I have to add one-metre flights of drill using mechanic's tools that are best used in warm conditions.

Once we get through the ice the actual fishing is relatively easy, provided we're at the right depth. Sometimes we even have camp chairs. Most of the time, though, it's backs into the wind staring down a small hole that constantly needs clearing with a tiny dip net because of accumulating ice. I was taught to fish with hand lines but soon shifted to little rods, and soft bait and thread instead of self-tangling

Victoria Metcalf (left) working with a field assistant to create a fishing hole through the sea ice using a Jiffy drill.

nylon. I have thought hard about ways to streamline the process, even purchasing a custom-made Antarctic rod.

Perhaps the ease of the catch is why there are always takers for my fishing trips. I went to Antarctica as a scientist and have returned as some kind of fishing guide, adept at untangling lines, finding the right piece of bait, and helping my new fisher people make sure they have their lines at the right depth.

The most abundant bottom-dwelling fish in shallow waters are the bernachs. Their diminutive and more sizeable cousins, *T. pennellii* and *T. hansoni* respectively, are often caught too. Rarer finds, requiring different techniques and/or locations, include the large predatory dragonfish *Gymnodraco acuticeps*, which comes complete with an ominous fanged lower jaw, and *T. newnesi*, an interesting species that appears to be undergoing a shift to a more pelagic lifestyle up the water column. Borchs are caught elsewhere, offshore over deeper water, with lines dangling just below the ice.

A shift northwards to slightly warmer climes in the Ross Sea leads to both familiar and unfamiliar catches. Bernachs, sometimes described as Ross Sea rats, are everywhere here. *T. eulepidotus*, pink and grey-blue in colour and taking the niche of borchs, are abundant in Terra Nova Bay. This is also a spawning ground for the ecologically pivotal *Pleuragramma antarcticum*, or silverfish, a pelagic food source for many other members of the Ross Sea's food web; we drill holes and scoop out ice slush to find its eggs. Silverfish are the key food species in the Ross Sea, in contrast to krill's importance in other regions of the Antarctica.

Down on the bottom lurks the large icefish *Chionodraco hamatus*. With its reptilian-like head and semi-transparent body – you can see its brain and the connections to its eyes – it is perhaps the most fascinating of fish. When it is dissected you can see its pale organs and its milky-white blood, devoid of red cells.

One fish I have never caught – rod and line would be insufficient – is the deep-water-dwelling Antarctic toothfish. However, I have seen those caught by Art DeVries and been gobsmacked by their size – an average fish casts a shadow over a fisherman when they are side by side, and specimens of over 160 kilograms have been caught. During my first visit to the ice, in an absolute one-off, I visited a fishing boat and afterwards consumed some of this fish's highly prized and delicate flesh.

While Antarctica's land is protected the protection does not extend to the marine environment, leaving it vulnerable to exploitation. The high levels of dissolved oxygen due to the low water temperatures support a richness of life in Antarctic waters: because of the planktonic algal growth that occurs on the sea ice and in the ocean each summer, the region can be considered the biggest supermarket on the globe for primary production, the process whereby energy is derived typically from photosynthesis. What happens down here matters to everywhere else on this planet.

The fecundity of the seas makes them prime targets for fishing activities. The Ross Sea has been described as one of the most pristine and untouched oceanic regions on Earth. There are many species here that are found nowhere else, from the microscopic through to the well-known charismatic megafauna. The abundant schools of small silverfish play centre stage in the food web, while toothfish take the place of sharks as apex predators. Humans aren't the only animals to consider toothfish a delicacy – they are a favourite food of Weddell seals and killer whales.

Fishing activities are just one of many threats to this unique ecology. Pollution, both historic and current, is currently quite small-scale but will grow as climate change accelerates, bringing with it long-range atmospheric and oceanic transfer of pollutants. One of the biggest threats is ocean acidification, caused when increased amounts of carbon dioxide gas in the atmosphere are absorbed by the sea, as though by a giant sponge. The world's oceans have already seen an average global decrease of 0.1 pH units since the start of the industrial age. This may sound trivial but it amounts to a thirty percent change in the amount of hydrogen ions.

Adding warming seas into the equation means a complicated future for marine life, especially when ice shelves and sea ice provide vital habitat. Current collapses of ice shelves are exceeding predicted rates. And new threats may appear. With increasing pressure on resources, we may see a stronger push for biological and mineral exploitation in Antarctic waters.

Spending many hours a day fishing outdoors in Antarctica has left me feeling deeply protective of, and concerned about, the fate of an ecosystem with such enormous value, both as a wilderness and as a habitat for wildlife found nowhere else. These remarkable creatures may be ill placed to adapt to warmer seas. The icefish family are particularly vulnerable because they lack the oxygen-carrying molecules haemoglobin and, in some cases, myoglobin. As temperatures increase, oxygen solubility decreases and these fish will struggle without an oxygen carrier. Icefish can therefore be considered a flagship species for investigating climate change impacts.

The more we learn the more we realise we don't know. Recently glaciologists drilling through the Ross Ice Shelf discovered, when they put a remotely operated SCINI vehicle down a hole, that there were fish living in a seemingly barren landscape under 740 metres of ice, 850 kilometres south of any sunlight that might drive primary production by photosynthesising phytoplankton. These ice-shelf fish inhabit sea water only ten metres deep in between ice above and rocky sea floor below. The scientists' drilling spot, at what is known as the Whillans Ice Stream, is where sea meets shore at the back of the ice shelf, and also where a glacier transitions from land to water – a place known as

The strange sight of a fisherman, TV presenter Clarke Gayford, at an ice hole at Cape Evans.

a grounding zone. These fish appear to be another as yet undescribed notothenioid fish species. Other kinds of fish were spotted too. What else might lurk under the ice shelf in this other-worldly realm the size of France? What else might be threatened?

"Fishing in Antarctica" by Victoria Metcalf is first published in this volume.

Ice-strengthened deep-water research vessel *Tangaroa* in pack ice in the Ross Sea, Antarctica.

Sea ice and polynyas

In February 2013 marine geologist Helen Bostock was part
of a twenty-two-member science team on a six-week New
Zealand-Australian-French voyage to Antarctica. Travelling on
RV *Tangaroa* – a research vessel run by NIWA, New Zealand's
National Institute of Water and Atmospheric Research – their
mission was oceanographic research in a transect across the
Southern Ocean from Wellington to the edge of the Antarctic
continent. The ultimate goal was to reach the Mertz Polynya
region of East Antarctica.

As a marine geologist, I have spent many months at sea on research
ships, but my first experience of seasickness was in the Southern Ocean
south of New Zealand. The fifty-knot south-westerly winds caused six-
metre swells, with a second swell coming in from the east. The large
seas saw most of the scientists confined to our cabins. After a couple
of miserable, debilitating days we emerged with our sea legs and got
to work setting up and testing equipment and getting to know each
other as we steamed south towards Antarctica.

The first voyage to the Mertz Polynya region had been more than a
hundred years before, led by the Australian geologist Douglas Mawson.
In 1911 thirty-five men, primarily young graduates from Australian
and New Zealand universities, set sail on the *Aurora* from Hobart in
Tasmania. Their main objectives were to explore the geography of the
Antarctic region, collect biological and geological samples, and make
oceanographic, meteorological and magnetic observations.

On their approach to the continent they encountered dense sea ice
and spent several frustrating weeks exploring potential leads through
the ice. They eventually reached the coast early in 1912 and built a

series of wooden huts – living quarters for men and dogs, and shelter for astronomical and meterological instruments – at two locations. They then spent a long winter trapped inside the huts by the raging storms that characterise the region.

In late November, when the weather improved, several teams set out to explore and make scientific observations. Mawson named many of the main geographic features after members of the expedition. The Mertz Glacier and Polynya – where we were heading – were named for Xavier Mertz, who died while he, Mawson and Belgrave Ninnis were returning from exploring the region east of Commonwealth Bay. Despite their relatively inadequate gear (at least compared to modern standards), the terrible weather conditions, and the deaths on the ice of both Mertz and Ninnis, Mawson and his team made many scientific discoveries, which have been built on by subsequent expeditions.

However, Antarctica is not static. The aim of our voyage was to study ongoing oceanographic changes in the region, with a focus on any alterations to the formation of Antarctic bottom water. This cold, dense, salty water forms an important part of the thermohaline-driven global ocean circulation. Driven by heat and salinity (affecting the density), the process not only controls the formation of large ocean water masses and currents but has a major influence on the global climate system.

Antarctic bottom water is formed in polynyas: the Russian word describes an open area of water surrounded on all sides by ice. Polynyas are formed by strong katabatic winds that drain high-density air from the Antarctic continent down towards the coast. When these winds, which can reach 300 kilometres an hour, reach sea level they blow the newly formed sea ice away from the coast. With the sea water now exposed to the extremely cold Antarctic air temperatures, more sea ice forms. Thus, polynyas are effective sea-ice factories.

Sea ice contains very little salt so its formation increases the salinity of the surrounding water. The remaining cold, very salty and dense sea water – called high-salinity shelf water – sinks to the sea floor and flows over the edge of the continental shelf, cascading down canyons on the Antarctic slope to the bottom of the Southern Ocean, where it forms the Antarctic bottom water. This cold dense water then flows north beyond the Southern Ocean, along abyssal plains at depths greater than four kilometres, and can be traced as far north as the equator.

Antarctic bottom water is primarily formed in three regions around Antarctica – the Weddell Sea, the Ross Sea, and the Mertz Polynya. In February 2010 the tongue of the Mertz Glacier, which extended approximately 100 kilometres from the coast, broke off after being rammed by B09B, a large iceberg that had calved from the Ross Ice Shelf in 1987 and moved into the Mertz region in 1992. The break resulted in a major change to the coastal area, altering the location and extent of the Mertz Polynya. This, in turn, influenced the distribution and formation of sea ice, and ultimately the location and rate of formation of the high-salinity shelf water that feeds the Antarctic bottom water. Oceanographers became concerned that shutting off the Mertz Polynya was significantly reducing the amount of bottom water produced, and that this would affect both global ocean circulation and climate.

To monitor the spatial formation of the shelf waters, oceanographers measure changes in salinity and temperature through the depth of the water column. This is done at different locations along the ship's voyage, to create a record of changing salinity and temperature around the region. Changes between seasons and years are also monitored using oceanographic instruments tethered to stationary mooring lines at different depths for a couple of years. These moorings allow oceanographers to monitor conditions in the water column throughout the year, not just during the summer season when the sea ice melts and ships can get in. This is especially important for understanding the formation of high-salinity shelf water as most of it is produced in winter when sea-ice formation is at maximum production due to the strong winds and extreme cold. The downside is that the data can be retrieved only when – and if – the moorings are recovered.

At a latitude of around 65°S we reached the edge of the sea-ice zone. Using daily satellite images to navigate, we skirted around the ice floes, searching for a way through the icebergs and sea ice to the polynya. The RV *Tangaroa* is not an icebreaker, only ice-strengthened, so it required considerable patience from the captain and the specialist ice pilot to ease the ship slowly through the leads in the ice floes.

In mid February several cloud-free satellite images revealed a large iceberg sitting directly over several of the main mooring sites. The area was surrounded by sea ice, with no obvious way through. The pictures showed a possible lead to the east so we steamed towards this to execute plan B – namely, retrieve another mooring close to what remained of the Mertz Glacier tongue. The satellite images turned out to be deceptive: there was a lot of sea ice. After several days making little progress we abandoned plan B.

Plan C was to work along the edge of the sea ice, leaving us with the option of sailing south to retrieve the moorings if and when a lead opened up. For the next week we took oceanographic measurements along the continental slope, looking for evidence of dense, high-salinity shelf waters flowing off the shelf and down the canyons into the deep ocean. Using a gravity coring system we also retrieved a series of sediment cores between four and six metres long from the ocean floor. These will be used to assess changes in the formation of Antarctic bottom water over longer time periods. Sediment cores can be useful climate archives, as mud and the remains of microscopic organisms accumulate on the sea floor. Variations in the species of organism, the chemistry of the microscopic skeletons, and the proportion of mud within a core indicate a change in environment. The sediments go back from hundreds to thousands of years, providing information about the natural variability of the region beyond the historical observations.

Finally a large lead opened up in the sea ice. It was much further east than the moorings we wanted to collect, but in a region from which little previous data had been collected. We headed along the edge of the sea ice, overlying the continental slope, and steamed south on to the continental shelf, getting as far as the Antarctic Circle at 66.6°S. It was now late February and air temperatures were dropping as low as −12°C. This made work on deck much harder. Any water left on the instruments froze quickly, blocking up sensors and preventing them functioning correctly. We also needed to be more careful to make sure our hands and faces didn't get frostnip or frostbite when we were working outside.

We collected some interesting data on the continental shelf. We found no evidence of dense, salty, cold water being formed in this region east of the Mertz Polynya, but we did find signs of high biological productivity. We collected large amounts of phytoplankton – the microscopic plants that form the base of the food chain – in our plankton nets, and the chemical data showed a large drawdown in carbon dioxide used in photosynthesis. This high phytoplankton productivity means the coastal regions around Antarctica are large sinks for carbon dioxide. Underwater video also showed abundant biodiversity on the sea floor.

By this time it was early March and pancake ice was beginning to form on the sea surface, the result of the accumulation of microscopic frazil ice. The pancakes grow as waves splash over them and freeze, and as more water freezes on to the bottom of them. Eventually they grow together to form ice floes. The captain and ice pilot were becoming increasingly uneasy about pushing through the rapidly forming pancake ice, worried the ship might get stuck and have to wait for an icebreaker to break it out. The captain ordered retreat. It was the end of the summer season in Antarctica and time to return to New Zealand, albeit without the oceanographic moorings.

The extent of sea ice around Antarctica in that late summer of 2013 was exceptional, the second largest since satellite records began in the 1980s. Why has the extent of Antarctic summer sea ice become greater when the extent of Arctic summer sea ice is rapidly declining? This is a puzzle scientists are trying to solve. Some say the ozone hole over Antarctica has resulted in a speeding up of the winds and a cooling over the southern continent. Others have suggested the melting of Antarctic glaciers has resulted in a cold freshwater surface lens, buffering the sea ice from the warming ocean waters below. Or is it just natural variability?

It will be important for today's scientists to continue monitoring the region over the next few decades so scientists of the future have data to help answer these and many other questions about the changing oceans around Antarctica. These regions may seem remote but changes there are likely to have a significant impact on global climate.

> Following the RV *Tangaroa*'s 2013 voyage, Australian and French researchers continued their efforts to retrieve the moorings. In January 2015 the crew of an Australian icebreaker, *Aurora Australis*, retrieved one French and two Australian moorings from the Mertz region. A third Australian mooring was located but retrieving it was deemed too dangerous: it was straddled by towering ice cliffs and overlain by ice floes.

Adapted from a series of blogs written by Helen Bostock with input from many of the voyage's participants, and published in *Field Work – Scientists on Expedition* at sciblogs.co.nz, 2013.

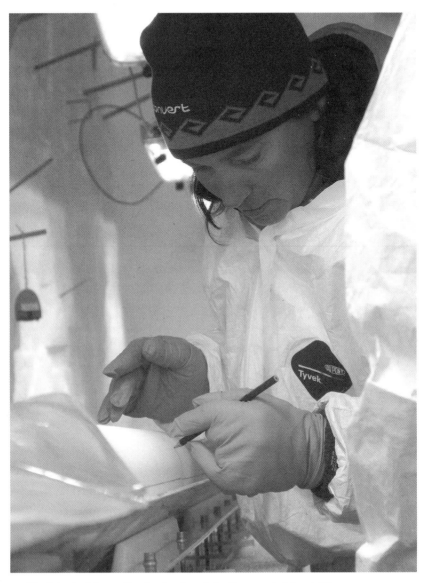

Glaciologist Nancy Bertler, leader of the Roosevelt Island Climate Evolution project, marks up a sample at a special field laboratory on Roosevelt Island, at the northern tip of the Ross Ice Shelf.

A page from the ice diary

In 2009 the Andrill marine sediment drilling project revealed that three to five million years ago – when atmospheric CO_2 concentrations were 400 parts per million, global air temperatures were three to four degrees Centigrade warmer than today, and the Southern Ocean was five to six degrees warmer – all of the West Antarctic Ice Sheet, all of the Greenland Ice Sheet, and the margins of the East Antarctic Ice Sheet collapsed. In response, global sea level rose to twenty metres higher than it is today. But Andrill didn't provide any information about how fast the ice sheets had collapsed.

In 2011, as atmospheric CO_2 levels approached 400 parts per million – a direct result of anthropogenic CO_2 emissions from burning fossil fuels – Nancy Bertler led an international team of drillers, geologists, glaciologists and support staff to establish a remote field camp at Roosevelt Island, which lies 683 kilometres from McMurdo Station, at the northern edge of the Ross Ice Shelf. With the collapse of the Ross Ice Shelf and the West Antarctic Ice Shelf now predicted responses to Earth's warming climate, the three-year ice-drilling project aimed to determine the speed of past changes, when the Ross Ice Shelf, and the West Antarctic Ice Shelf behind it, retreated. Did this occur over 50 or 500 years?

Bertler, a glaciologist who trained in Germany, the UK, New Zealand and the United States and now holds a joint position between Victoria University of Wellington's Antarctic Research Centre and GNS Science, describes the story behind this Roosevelt Island Climate Evolution (RICE) project.

The Antarctic ice sheets are very dynamic, advancing and retreating – or even collapsing – in response to global air and ocean temperatures.

The Andrill project showed us that three to five million years ago, in a warmer climate, the West Antarctic Ice Sheet and the margins of the East Antarctic Ice Sheet collapsed. But analysis of marine sediments is not precise enough to show how fast the ice sheets collapsed. For that, we need ice cores.

There are no ice core records that go back three to five million years, but there was another time when the Ross Ice Shelf and West Antarctica experienced dramatic changes, and that was after the last glacial maximum, about 20,000 years ago.

During this glacial maximum, global temperatures were about six degrees cooler than today, the West Antarctic Ice Sheet and the Ross Ice Shelf were much larger, and the sea level was lower. Then 20,000 years ago things started to warm up. First, the northern hemisphere ice sheets melted and the sea level rose by 120 metres. It was only when global sea levels stabilised, about 10,000 to 8,000 years ago, that the Ross Ice Shelf started retreating – by about a thousand kilometres – to where it is today. But we don't know how this ice shelf disintegrated. Was it a smooth, gradual retreat or did big chunks of ice break off? And when it went, how quickly did it go?

To investigate what happened over this warming period, we chose Roosevelt Island, which is at the northern tip of the floating Ross Ice Shelf. We anticipated we could drill an ice core here that was deep enough to take us back at least 30,000 years and perhaps a lot more. This ice core could give us information about past air temperatures, sea surface temperatures, changes in the extent of sea ice, and more. Because Roosevelt Island is so close to the edge of the Ross Ice Shelf, the site could also provide information about advances and retreats of the ice shelf over this time since the ice at Roosevelt Island would have thickened and thinned as the shelf grew or retreated. By looking at temperature changes in combination with the movement of the ice shelf, we could reconstruct how sensitive the ice shelf was to temperature changes.

To get to Roosevelt Island, a raised blob of ice sitting at the northern edge of the Ross Ice Shelf, we flew for two-and-a-half hours across the ice shelf from McMurdo Sound. At this very remote location – the mountains of East Antarctica are about 700 kilometres away and the West Antarctica icefalls almost 200 kilometres away – the view is white in every direction. The northern edge of Roosevelt Island is exposed to the ocean, but with the frequent blizzards and regular fog rolling in from the ocean we couldn't see it.

Our first priority was to set up the camp – kitchen tent, bathroom tent and sleeping tents. Next priority was setting up our workspace, a 27-metre-long trench covered with a tent to protect our drilling rig and clean laboratory. Outside temperatures ranged from about minus 30°C early in the season to minus 5°C during the peak of the summer. In this dry environment, with our Antarctica New Zealand-issue layers of clothing, minus 15°C was the perfect working temperature. But even that was too warm for our ice cores. Ice cores are extremely sensitive – if they warm up higher than minus 18°C they start to deteriorate. So, for this remote Antarctic field camp, we had to take a freezer with us to keep these cores cold.

During each four-month field season we drilled for twenty hours a day, using a purpose-built drill that was light enough to be flown into the camp yet robust enough to drill down to 1,000 metres. Instead of making a hole the drill cut a ring, leaving the piece in the middle, the core, intact. Each time we sent the drill down on a wire it cut another two metres into the ice. We then broke the core and brought it up on a winch. At the surface we extracted the inner barrel that contained the core, and slid the ice out on to the processing table. The core-processing team fitted this core to the previous piece of core, cut this into one-metre-long pieces and packed these in ice core boxes in a storage cave off the main trench, ready to be flown back to Scott Base and then shipped back to New Zealand.

How can an ice core tell us so much about past climate? The answer is that the core is like a detailed diary. Every day it snows, a page is added to this diary. As well as snow conditions, the ice diary records wind-blown material such as marine silts, dust particles, volcanic eruption products, and even tiny organisms. And all the snow and particles that find their way to this place, which is incredibly clean, start to accumulate, layer by layer, or page by page. So when we drill though the ice we go back in time, and the ice – and what's trapped inside it – tells us the story of what was happening when the snow was falling.

Back in the ice-core processing laboratory in New Zealand, we analyse the ice itself, the air bubbles trapped in the ice, and the particles in the ice. First we melt a section from the pristine middle of the ice core. As the ice from the core is melted, we look at the composition of the water. The hydrogen atoms in the water molecules can tell us about the temperature when this snow was deposited. A deuterium atom is a stable isotope of hydrogen that has a proton and a neutron in the nucleus. The higher the proportion of these atoms to regular hydrogen atoms, which have only one proton in the nucleus, the warmer the temperature.

We also look at the chemistry of the water, including the abundance of sea salts (such as sodium and chloride) relative to terrestrial elements (such as magnesium, calcium, and potassium); this can tell us about storm tracks. While cyclones originating in the Southern Ocean deposit sea salt, the katabatic winds, those strong gravitational winds blowing from the heart of the Antarctic continent, carry terrestrial elements.

In addition, we can look at the concentration of sulphate to reconstruct the emissions of past volcanic eruptions, and the concentration of a chemical called methanesulfonate that is created by plankton in the ocean and provides a measure of the amount of phytoplankton in the Ross Sea.

As we melt the ice, any bubbles separate out from the water. We capture these bubbles, release the atmospheric gases, and analyse the composition of this ancient air. We're particularly interested in the proportion of greenhouse gases such as CO_2 and methane.

Then we look at particles, which can also tell us about amount of phytoplankton, about southern hemisphere volcanic eruptions – which can have a cooling effect on climate – and about dust concentration and composition, which are really important for the climatic system. As dust is deposited in the ocean it provides nutrients to marine ecosystems, primary productivity increases, and the resulting plants gobble up atmospheric CO_2.

By analysing the ice and the air and particles it contains we can reconstruct conditions such as atmospheric circulation, temperature – how cold the winters were and how warm the summers – and the extent of sea ice. We can distinguish between the extent of the sea ice and a more extensive Ross Ice Shelf because in summer the sea ice retreats – leaving a distinctive annual signature – but the ice shelf does not. This provides a comprehensive view of past climate and the way the ice shelf and the ice sheets responded. Analysis of the data so far has revealed that the Ross Sea region has been extremely dynamic over the past 30,000 years. There have been large changes in ice volume, ocean temperature, sea ice extent and ultimately ice shelf retreat.

In January 2013, at the end of the second season of drilling, the RICE team hit bedrock after collecting 764 metres of ice core. Processing the samples at the GNS Science National Ice Core Laboratory involved an international team working for a total of four months in twelve-hour shifts and twenty-four-hour days, much of the time in a minus 36°C freezer.

"A page from the ice diary" by Nancy Bertler is first published in this volume.

A mobile fishing hut near Scott Base, Antarctica, where Regina Eisert and her fellow scientists fished for Antarctic toothfish, the Ross Sea's apex predator.

What do seals want for Christmas?

The Ross Sea is one of the world's most pristine marine habitats but these wild Antarctic waters are also host to a lucrative international fishery, led by ships from New Zealand, which sell the toothfish catch as "Antarctic cod" and "Chilean sea bass". Amid concerns about the impact of this commercial fishery and other factors on the ecosystem, New Zealand and the United States have proposed the Ross Sea be declared a marine protected area to balance "marine protection, sustainable fishing and science interests".

There's still a lot to learn about this ecosystem. "Our ability to predict or manage impacts is limited by a lack of knowledge," says German-born marine biologist Regina Eisert of the University of Canterbury's Gateway Antarctica. Eisert and her team have been studying the food requirements of killer whales, Weddell seals and Adélie penguins "to provide reference points for detecting future change and to identify what food resources are critical to these predators to allow responsible environmental stewardship of the Ross Sea". In the Antarctic summer of 2014–15 Eisert and her collaborators spent three months working out of Scott Base. In this series of posts first published on Antarctica New Zealand's website, she writes about her work with toothfish and seals.

November 12, 2014: A new kind of field assistant
The call comes over the public speaker system at Scott Base. "We have a toothfish. Who wants to help us carry it to the science building?" Within minutes people in orange and black jackets show up, ready to share in the excitement of science on the ice.

Our team – which includes Art DeVries (University of Illinois), Sophie Mormede and Steve Parker (both National Institute of Water and Atmospheric Research) and me – is fishing for Antarctica's largest fish. Toothfish can grow to more than two metres and weigh over 100 kilograms. This particular one weighs "only" thirty-three kilograms and measures 1.33 metres without its head or inners, which have been removed by our friendly field assistant.

We are here in Antarctica to solve a mystery. Art DeVries has been catching toothfish in McMurdo Sound since the 1960s; he discovered fish antifreeze and is the world's foremost authority on Antarctic toothfish. For several decades, he reliably caught toothfish on vertical set lines every year in McMurdo Sound but in 2003 the catch drastically declined. An obvious answer is that the fishery may be responsible for the empty hooks, yet the toothfish stock surveys carried out each year by scientists from NIWA (including Steve and Sophie) reveal no sign of a radical population decline.

We have decided the best way to figure out what is going on is to work together. After Scott Base staff have set up a fishing hut for us, Sophie, Steve and Art set lines through a hole in the sea ice every day, but without success. Within hours of setting the first line a large, sleek, beautifully marked Weddell seal showed up at our hut and started using our fishing hole to breathe and sleep, resting its chin on the edge of the ice. Bits of fish floating in the hole revealed that, while our hooks might stay empty, the seal had no trouble finding toothfish. And tonight our new field assistant leaves a toothfish for us, minus the head and guts, which are apparently the seals' favourite bits and eaten first. So we have got our first toothfish at last, just not quite the way we have planned. Next week we will try our luck at a new hole a little further away from helpful seals.

November 17, 2014: Gone fishing, not gone fish

Steve has identified a promising spot, a previously unexplored basin to the north-west more than 600 metres deep. Toothfish prefer deep water and the site is far away from seal colonies. After the Scott Base engineering team moved the hut to the new spot, Steve, Sophie and Art set their first line two days ago and hauled it in yesterday. They caught six enormous toothfish on the first line.

I gather a team of keen volunteers from Scott Base to pick up the catch. While we triumphantly escort the fish back to Scott Base, Steve and the others re-set the line. Was this just beginner's luck, or are there really lots of big toothfish? Today brings the answer: another six beautiful fish on the line, with lost hooks hinting that some big monsters may have gotten away. Not only is that a stunning success, the fish are uniformly large (1.25 to over 1.5 metres long), precisely the size that was thought to have disappeared from McMurdo Sound. The biggest fish so far weighs 50 kilograms and takes two people to move.

Once the fish are back at Scott Base, the work is not over. Steve and Sophie sample the fish and collect the otoliths, or "earstones", that are used to determine the fish's age. Even without looking at the otoliths, Steve can tell the fish are adults in good condition. Once the sampling is done, the fish are bagged up and put in the Scott Base science freezer to await their transport north. (I get many requests to turn our samples into sushi but these fish are strictly for science!)

Back in New Zealand I will analyse the toothfish for fat, protein, water, minerals and energy: my hypothesis is that toothfish are superior to other prey as food for seals and killer whales. Because they don't come with food labels we have to analyse them to know how many toothfish a whale has to eat to meet its daily requirements.

December 4, 2014: Seals, tides and an epic adventure

Today, Tom Arnold, the field assistant, Kate McKenzie, the science technician, Shaun Clarke, the helicopter pilot and I installed two survey cameras on top of Turtle Rock. For days we had been trying to reach the summit of the island, but poor visibility and hidden cracks in the sea ice had kept us away. The solution was simple: take one extremely experienced pilot, a highly motivated team, and drop in by helicopter.

Turtle Rock, a tiny island just north-west of Scott Base, is best described as a pile of volcanic scree that vaguely resembles a turtle when viewed from the north (especially if you have been looking at nothing but snow and ice for a few months). It splits the sea ice west of Hut Point Peninsula into a perennial tide crack, which attracts a small colony of Weddell seals: they use the crack to get in and out of the water. With the help of the Scott Base team I had been trying to set up a couple of automated time-lapse cameras on the summit of Turtle Rock to monitor the daily activity of the seals, but our attempts had been frustrated by bad weather.

But why would anyone want to spy on seals? While the seals around McMurdo are counted and tagged every season, nobody really knows how many Weddell seals there are in the Ross Sea, and knowing seal numbers is essential for detecting the effects of climate change, fishing and other changes on the sea. Weddell seals are one of the top predators and, as such, a sensitive indicator of the state of the ecosystem.

Counting all the seals in the Ross Sea is feasible only by using remote – aerial or satellite – surveillance. Weddell seals are ideal for this in that they are large, black, predictably congregate in breeding colonies during spring, and stand out well against the ice. The only problem is that seals disappear beneath the ice to go diving: more than ninety percent of Weddell seals in a colony will be under the ice and invisible during the course of a given day. To get accurate numbers you need to either count the seals when they are all hauled out on the surface, or correct the counts to include the seals that are away diving.

Early studies reported that the seals in McMurdo were all hauled out at four p.m. every day. However, data my team and I had collected on a previous expedition suggested their daily schedule might be entrained, or influenced, by the tides. This matched research by German scientists, who found a correlation between seal diving and tides.

To get a solid algorithm that could use the tidal cycle to correct satellite data, I urgently needed more data. With bad weather keeping us from the island, time to deploy the survey cameras was running out – soon the weakening sea ice would break out and put Turtle Rock out of reach. Luckily IBR, the Scott Base helicopter, and its pilot Shaun came to the rescue. The decision to drop us in by helicopter to the top of Turtle Rock is an example of the kind of easy-going competence and can-do attitude that makes the New Zealand Antarctic programme what it is – effective, flexible, and punching well above its weight in terms of science supported to dollars spent.

Tomorrow Tom and I are going back to check on the cameras. Within a week or two we should have a better idea of how seals keep time. I'll keep you posted!

December 31, 2014: What do seals want for Christmas?

It is the end of the year, and the sea ice in McMurdo Sound is cracking and thinning under the bright summer sun. The blanket of ice that has covered up the ocean all winter long is gradually being pulled back, and the life-giving sunlight is turning the sea water a vivid shade of green. Weddell seals are moving away from the colonies where they have been nursing their pups, taking advantage of the cracks in the ice to occupy new areas with access to the sea. One of the traditional summer holiday hang-outs for Weddell seals is in the pressure ridges near Scott Base, where the sea ice is squeezed into folds and troughs between Hut Point Peninsula and the Ross Ice Shelf. This year the pressure ridges have formed right in front of Scott Base, affording us a close-up view of the seals sunbathing on the ice and wallowing in the melt pools.

A Weddell seal displays the unusual behaviour of eating a toothfish on the ice. It had previously been believed that seals ate only in the water.

At Scott Base, Christmas is celebrated with a lovely meal and Secret Santa, with everyone on base receiving a handmade gift from an anonymous giver. Because recipients are known to givers well in advance, a lot of thought goes into what each person would like, and many people receive the best presents ever. Mine was awesome, and a close second only to the horse I got for Christmas when I was twelve.

I had never given much thought to what seals want for Christmas. Over the years that people have been in McMurdo Sound there have been a number of observations of Weddell seals catching large Antarctic toothfish, often using fishing holes made by scientists. All reports since the first was published in 1969 describe the same behaviour: the seal arrives at the hole with a live toothfish firmly grasped by the lower lip, and holds the fish above the surface where it can't breathe. Once the fish is weakened by lack of oxygen, the seal tears it to pieces, first breaking off the head, then removing the guts, and finally devouring the rich white flesh. Seals don't bother to eat the tail end, and may stash a half-eaten fish under the ice for later. (Toothfish are so high in fat that they float once the head has been removed.)

But this year we saw something nobody had ever seen before. Paul Kelly and Bill Strickland, two volunteers seconded to Scott Base by the New Zealand Antarctic Society, were going for a Christmas Day walk in the pressure ridges when they spotted a seal with an enormous toothfish on the sea ice, resting its head contentedly on its catch, looking like a happy Labrador dog. The accepted wisdom is that seals eat only in the water, and there are no previous records of a Weddell seal bringing a toothfish out on to the sea ice. But this one did – and got exactly what it wanted for Christmas.

Adapted from "Update from the Ice" by Regina Eisert: antarcticanz.govt.nz, November 12, 2014.

A king crab found on the Antarctic sea floor, 2015. Until 2003 there were no crabs on Antarctica's continental shelf, but warming sea temperatures have made it possible for these predators to invade and survive there and threaten the unique inhabitants.

March of the king crabs

Kathryn Smith is a research biologist at the Florida Institute of Technology. Here she describes her current research project, in which she is examining how a rapidly warming ocean is enabling predators to invade the unique communities that live on the Antarctic sea floor.

Hundreds of metres below the surface of the freezing ocean surrounding Antarctica the sea floor is teeming with life. The animals living here have no idea that an army is on the brink of invading their tranquil environment. The army is composed of king crabs. Until 2003, there were no crabs in this fragile Antarctic ecosystem. Now, driven by warming waters, their arrival heralds a major upset.

The unique communities living on the continental shelf off Antarctica are found in no other place on Earth. Delicate brittle stars, beautiful sea stars, vibrant sea lilies and giant sea spiders are among the spectacular inhabitants. The animals live side by side, with almost no predators to upset the balance.

For millions of years, the cold water temperatures in the Antarctic have stopped most predators from surviving in this harsh environment, but this is rapidly changing. Climate change is increasing temperatures across our planet, and the Antarctic is no exception. Sea temperatures are rising at a faster rate here than almost anywhere else.

With the increasing temperatures come new residents. Animals that have been absent from the continental shelf around Antarctica for millions of years are quickly returning.

A prime example are crabs. In every part of the world except Antarctica, crabs are one of the major predators in sea-floor communities. Their strong crushing claws are deadly to snails, brittle stars and other slow-moving animals. On the continental slope and continental shelf surrounding Antarctica, icy water has kept the crabs away. Crabs naturally take up magnesium into their blood from sea water and can normally control the levels present. However, at very low temperatures they lose this ability and the magnesium builds up, acting like an anaesthetic and eventually causing the crab to die.

Water temperatures around Antarctica are now warming to levels that crabs can tolerate. Although they and other predators have long been absent from the area, one group, king crabs, have been living in the neighbouring deep ocean, where water temperatures have historically been warmer. Now, as shelf temperatures increase, they are beginning to move. In 2003, king crabs were seen on the continental slope off Antarctica for the first time. Since then growing numbers have been reported. The crabs are seemingly marching up the continental slope and towards the continental shelf with nothing to stop them.

If king crabs move on to the shelf they will be presented with a smorgasbord of invertebrates. King crabs do not care much what they eat. Any animal that falls into their path makes a delicious treat. In the Antarctic the native inhabitants are particularly at risk. In other parts of the world, animals living on the sea floor have thick shells or hard skeletons to protect them against predators such as crabs, but in Antarctica they have evolved without any major predators for millions of years, and so their defences are limited. Their very thin shells, soft bodies and light skeletons make them an easy target for the rapidly approaching king crabs.

When these crabs arrive, they are very likely to have a huge impact on the unique inhabitants. These defenceless animals may well become yet another casualty of climate change.

King crabs are not alone in their invasion of Antarctica. Several non-native species of plants and invertebrates have also made their way

A yellow sea lily (top) and brittle stars on the continental shelf. The brittle stars' thin shells, soft bodies and light skeletons make them and other communities that have existed here for millions of years an easy target for king crabs.

south and are in the process of settling into the cold barren landscape. Although such invasions are rare, they occur most commonly around the western Antarctic Peninsula where tourism is high.

Most shallow-water and land-based alien species arrive with the help of humans. Each year tens of thousands of people visit the Antarctic for tourism or science. In the 2014/15 season more than 46,000 arrived by cruise ship alone. The region is also seeing a growth in commercial fishing, both legal and illegal. Shallow-water marine animals and plants can be carried on the hull or in the ballast water of a fishing boat, cruise ship or research vessel. For example, a species of shallow-water crab has recently been found frequenting the coastline of Deception Island. This island is just off the western Antarctic Peninsula and one of the most popular stops for cruise ships and other vessels. It is likely it was one of these boats that carried the crab to Deception Island from southern Chile, although it may also have been borne by ocean currents or on floating debris.

Plants and other terrestrial species usually arrive in Antarctica attached to the footwear, clothing or equipment of tourists or scientists. Seeds, buds, suckers and spores of plants, including moss, have been found on many pieces of expedition equipment. Alien species of plant can also arrive aboard aircraft, or be carried by wind or migrating birds.

Both on land and in the water, alien species can wreak havoc among the vulnerable native ecosystems. Because of this, one of the aims of the Antarctic Treaty is to protect the continent from such invaders. Where possible, non-native species are removed. Most alien plants that have been found growing around the western Antarctic Peninsula have now been successfully eradicated, stopping them from spreading. However, invertebrate species such as flies, mites and worms are more difficult to remove because of their mobility. Large areas of habitat would need to be destroyed, and as a result to date no attempts have been made to wipe them out.

As we've seen, climate change is making conditions increasingly favourable for alien species in Antarctica – in shallow water and on land, as well as on the sea floor. The rising temperatures make it easier and easier for new species arriving here to stay. Consequently, it is no surprise that biological invasions are now thought to be the biggest threat to the continent's conservation. As Antarctica continues to warm and ship traffic around it increases, the threat of invasions will only rise. Human activities may one day make the unique ecosystems of Antarctica, which have existed in isolation for millions of years, disappear completely.

Adapted from "The March of the King Crabs: A Warning from Antarctica" by Kathryn Smith: theconversation.com, July 21, 2015.

Glossary of scientific
and nautical terms

ablation Removal of snow, ice or water from a glacier or snowfield by sublimation, melting, evaporation or wind.

addled (of egg) Rotten, not producing a chick.

ammonite Extinct aquatic mollusc. Ammonites were abundant in the Mesozoic era (252–66 million years ago) and are commonly found as fossils in rock strata of that time.

anchor chain (balloon) Chain to control a balloon's height. The chain is looped from the balloon through an anchor in the ground, then attached to a winch.

anemometer Instrument for measuring wind speed.

aragonite Form of calcium carbonate from which the shells of many molluscs are formed.

archaeocyaths Extinct group of reef-building marine organisms from the Lower Cambrian period, which started 540 million years ago.

aspiration psychrometer Instrument for measuring atmospheric temperature and humidity, consisting of wet and dry bulb thermometers

athwartships Across a ship, from one side to the other.

aurora australis Luminous coloured atmospheric phenomenon seen in the night sky in high southern latitudes, caused by interaction of atoms and molecules in the upper atmosphere with charged particles from the sun.

Aztec Siltstone Geological formation consisting of layers of sandstone and red and green shaly siltstone from the Devonian period, found in areas of Australia and Antarctica.

barograph Barometer that gives a continuous recording of air pressure.

benthic life Life that occurs at the bottom of a body of water.

berg Iceberg. A large piece of floating ice that has broken off a glacier.

bergy bit Chunk of freshwater ice; smaller than an iceberg, but bigger than a **growler**.

brash ice Fragments of ice, less than 2 m in diameter, that have broken off icebergs or ice floes, often covering large areas but not hindering navigation.

Brunton transit instrument Type of precision compass first used in the late nineteenth century.

Canopus Brightest star in the constellation Carina; the third brightest star in the sky after the sun and Sirius.

carronade Short, light cannon used on Royal Navy ships from the 1770s to the 1850s.

chlorofluorocarbons (CFCs) Ozone-depleting chemicals consisting of carbon, chlorine and fluorine.

chondrite Non-metallic stony meteorite; the first kind of meteorite ever found in Antarctica.

chromatograph Apparatus for separating out and measuring the concentrations of different chemicals in a mixture, e.g., a sample of soil or water.

cirrus cloud Form of white wispy cloud, typically at high altitude.

col High pass or saddle in a ridge. It may mark the line of a former stream valley or glacier, and so provide evidence of an early stage in the development of the landscape.

confervae Any simple filamentous green algae.

coruscation Glittering or sparkling quality.

cosmic rays High energy charged particles, travelling at speeds close to the speed of light, which reach Earth from space.

Cretaceous period Youngest period of the Mesozoic era and the last portion of the dinosaur age, from c. 145–66 million years ago.

cumulus cloud Dense, isolated and clearly defined cloud with bulges or domes and a flattened, darker base. Most common at low altitudes.

cyanobacteria Phyllum consisting of two groups of photosynthetic eubacteria, the most common being blue-green bacteria, also known as blue-green algae. Very dense microbial communities can form large coloured mats in lakes and rivers.

dead reckoning Process of estimating the position of a ship based on its previous position and its course and speed over a known interval of time.

deionised water Water from which ions have been removed. Used for many purposes as an alternative to distilled water.

Devonian period Period of the Palaeozoic era stretching approximately from c. 419–358 million years ago.

dolerite Dark-coloured igneous rock commonly found in shallow level intrusions such as dykes, sills or plugs.

Eppendorf pipettes Brand of pipette, a slender tube with a bulb at the end, used for transferring or measuring small quantities of liquid.

Esquimaux Early alternative spelling of Eskimo, people inhabiting the northern regions of Canada, Greenland, Alaska and East Siberia; also known as Inuit.

fathom Unit of measurement equal to 1.83 m.

geophone Instrument used to detect the arrival of seismic waves by transforming the ground motion into an electrical voltage.

glacier tongue Glacial ice extending seaward from shore, often the source of icebergs. Also used to describe the terminal portions of valley glaciers, which are often shaped like a tongue.

glaciologist Expert in the study of glaciers, or more generally ice and snow, and their physical properties.

Gondwana Former supercontinent from which South America, Africa, India, Australia, New Zealand, and Antarctica are derived. Its break-up began in the Early Jurassic period, c. 180 million years ago.

growler Piece of ice, smaller than a bergy bit, and almost submerged beneath the sea's surface and therefore dangerous to ships.

halon Gaseous carbon compound containing bromine and other halogen atoms (chlorine, fluorine or iodine) and used in fire extinguishers. Halons are both atmospheric ozone depleters and greenhouse gases.

hemocyte Blood cell of an invertebrate (e.g. krill).

hoosh Thick soupy stew eaten by heroic age Antarctic explorers; usually a mixture of pemmican, crumbled biscuits and water or tea.

hygrometer Instrument for measuring atmospheric humidity.

ice island Old term for tabular icebergs, which can be several hundred square kilometres in area, with steep sides and a flat top. Sometimes also used to mean an island covered in ice.

Jiffy drill Brand of motorised, hand-held ice drill.

Jurassic period Middle period of Mesozoic era, approximately 201–145 million years ago.

katabatic winds Gravity-driven winds caused by movement of dense cold air down slopes; when travelling off the Polar Plateau, these winds can reach 300 km an hour.

Kemmerer Metal tube, with stoppers at each end, used to sample water, e.g. from a lake. The stoppers close when the sampler is in the water, and a metal cylinder called a messenger is dropped down the line holding the sampler.

leads Transient areas of open water between expanses of sea ice.

leptolyngbya Genus of cyanobacteria.

limnology Study of inland water bodies (e.g. lakes, rivers, groundwater).

lithosphere Earth's crust and upper region of the mantle, which is fragmented into tectonic plates.

Lloyd-Creak dip circle Instrument used to measure the angle between Earth's magnetic field and the horizon, with a second needle to measure the relative intensity of the magnetic field.

lycophytes Spore-bearing vascular plants, similar to ferns, that first emerged c. 410 million years ago.

lycopod A type of lycophyte, reaching upwards of 20 m.

magnetic deviation Deviation of a compass needle from true north, measured in degrees, defined as positive towards the east and negative towards the west.

magnetic dip Angle between the direction of Earth's magnetic field and the horizon.

magnetic field (of Earth) Magnetic field that extends from Earth's interior and out above the atmosphere, where it meets the solar wind. Magnetic field lines enter Earth at the north and south magnetic poles.

magnetic meridian Imaginary circle on Earth's surface that passes through the north and south magnetic poles. A compass needle on Earth's surface influenced only by Earth's magnetic field will come to rest along a magnetic meridian.

magnetic pole Part of Earth's magnetic field where lines of the field align vertically and enter Earth; compasses point to magnetic, not geographic, poles. The magnetic poles are not fixed at a geographical location but "wander" over time. When first located in 1908, the south magnetic pole was in northern Victoria Land, the area of Antarctica south of New Zealand. It is now well offshore, and north of the Antarctic Circle.

mesa Flat-topped hill normally underlain by near-horizontally bedded sediments.

meteorite Extraterrestrial mass that reaches Earth from outer space without burning up in the atmosphere.

meteoriticist Expert in the study of meteorites and other extraterrestrial materials.

microclimate Local climate of a small area or particular habitat.

moraine Rocks, boulders and debris carried and deposited by a glacier or ice sheet.

MRI Magnetic resonance imaging, a technique that uses a magnetic field and radio waves to create detailed images of the organs and tissues within a body.

mukluks Thick-soled boots for wearing in cold, icy conditions, with removable liners that can be taken out and dried.

neutrino Elementary particle with zero charge and tiny mass; it has a weak interaction with matter and is difficult to detect.

névé Young granular snow that has been partially melted, refrozen and compacted.

ocean acidification Process whereby increased amounts of carbon dioxide absorbed by sea water lead to increased ocean acidity.

ozone spectrophotometer Instrument that uses ultraviolet light from the sun to measure the amount of ozone in the atmosphere.

palaeontologist Expert on the study of life in the geological past.

parhelion Atmospheric optical phenomenon, caused by floating ice crystals in the air, which produces luminous spots on either side of the sun. Also called a mock sun.

pemmican High-energy mixture of dried meat, fat and berries that was eaten by many heroic age Antarctic explorers; see also hoosh.

photometer Instrument for measuring brightness of stars or other objects.

piedmont ice Bulb-like lobes of ice formed when steep valley glaciers spill and spread out into relatively flat plains.

Pisten Bully Brand of vehicle designed for traversing snow and steep gradients.

plate tectonics Concept that Earth's lithosphere consists of a number of distinct oceanic and continental plates, whose relative positions alter over the course of geological time.

pleopod Abdominal appendage on crustaceans.

polar amplification Phenomenon whereby changes to Earth's climate occur more extremely at the poles than is the average for the planet.

polar vortex Body of cold high-pressure air, encircled by westerly winds, formed during the long polar night and dissipating each spring.

polynya Large area of open water surrounded on all sides by sea ice, created by a combination of upwelling of deep, relatively warmer water and persistent winds, especially katabatic winds, or currents. Highly productive areas important for marine life.

proteridophytes Broad collective term for early terrestrial, leafless, spore-bearing, vascular plants.

Sargasso weed Brown seaweed found floating freely in tropical waters, particularly in the Sargasso Sea.

sastrugi Ridges of snow, up to 3 m high and created by strong winds, which harden into ice and make travelling difficult.

Secchi disk Black and white disk with a line and sinker attached, used to determine water clarity.

Silurian period Palaeozoic period during which the first life appeared on land, c. 444–419 million years ago.

Six's thermometer Thermometer that can record maximum and minimum temperatures over a period of time.

Ski-Doo Brand of snowmobile, or motorised over-snow vehicle, with skis at the front and tracks at the back; also called skidoo.

sling psychrometer Instrument, consisting of wet and dry bulb thermometers, for measuring atmospheric humidity; it is attached to a handle and spun in the air.

Southern Cross Constellation in the form of a cross visible throughout the year in the southern hemisphere.

stearine Soft odourless solid that can be extracted from many natural fats.

Tertiary period Former name for the first part of the Cenozoic era, c. 66–3 million years ago.

theodolite compass Surveying instrument for measuring horizontal and vertical angles.

thermograph Instrument for recording temperature changes. Expansions and contractions of alcohol in a tube drive a pen up and down over a rotating sheet of paper.

variation (magnetic) Angle between magnetic north and true north.

ventifact Stone, often found in Antarctica's Dry Valleys, with flat and often highly polished surfaces formed by sandblasting in strong winds.

vertical polarisation Process where an antenna's electrical field runs perpendicular to Earth's surface. Used for detecting neutrinos in ice.

Sources and copyright

Cook circumnavigates the continent
James Cook, *A Voyage Towards the South Pole and Round the World*, volume one: W. Strahan and T. Cadell, London, 1777: pp. 251–258, 264–270.

Pickled penguins and barrels of ice
Frank Debenham, ed., *The Voyage of Captain Bellingshausen to the Antarctic Seas 1819–1821*, volume one: Hakluyt Society, London, 1945, pp. 111–113, 122–124.

Drinking wine in Adélie Land
George Murray, ed., "The Journal of M.J. Dumont-D'Urville" in *The Antarctic Manual for the Use of the Expedition of 1901*: Royal Geographical Society, London, 1901, pp. 447–450.

Wilkes among the ice islands
Charles Wilkes, *Narrative of the United States Exploring Expedition, During the years 1838, 1839, 1840, 1841, 1842*, volume one: Ingram, Cooke, and Co, London, 1852, pp. 263–269, 273.

Fire and ice
James Clark Ross, *A Voyage of Discovery and Research in the Southern and Antarctic Regions During the Years 1839–43*, volume one: John Murray, London, 1847, pp. 216–221.

The polar captain's wife
Chris Orsman, *The Lakes of Mars*, Auckland University Press, Auckland, 2008, p. 12.

In a sleeping bag beneath the aurora australis
Frederick A. Cook, *Through the First Antarctic Night 1898–1899*: DoubleDay, Page & Company, New York, 1909, pp. 210–215.

The art and science of sledge travel
Admiral Sir F. Leopold McClintock, "On Arctic Sledge-Travelling" in *The Antarctic Manual for the Use of the Expedition of 1901*, edited by George Murray: Royal Geographical Society, London, 1901, pp. 294–302.

Sources and copyright

Skinned penguins and bloody seals
Edward Wilson, *Diary of the Discovery Expedition to the Antarctic Regions, 1901–04*: Blandford, London, 1966, pp. 88–96. Edited from original manuscript in Scott Polar Research Institute by Ann Savours, 1966. Excerpts by permission of Scott Polar Research Institute, University of Cambridge.

Drygalski's balloon ascent
Erich von Drygalski, *The Southern Ice-Continent: The German South Polar Expedition aboard the Gauss 1901–1903*, translated by M.M. Raraty: Bluntisham Books / Erskine Press, 1989, pp. 157–158.

Luxuriant vegetation and extensive coasts
Otto Nordenskjöld and Johan Gunnar Andersson, *Antarctica: Or Two Years Amongst the Ice of the South Pole*: C. Hurst & Company, London, 1977, pp. 244–252.

Penguin eggs and fried seal
Otto Nordenskjöld and Johan Gunnar Andersson, *Antarctica: Or Two Years Amongst the Ice of the South Pole*: C. Hurst & Company, London, 1977, pp. 406, 449–459.

Hunting the south magnetic pole
Edgeworth David, "An Account of the First Journey to the South Magnetic Pole" in *The Heart of the Antarctic*, edited by Ernest Shackleton, volume two: Heinemann, 1909, pp. 170–182.

Nematodes, rotifers, water bears and mites
James Murray, "Part II: Microscopic Life at Cape Royds" in *British Antarctic Expedition 1907–9, Reports on the Scientific Investigations*, volume one: Heinemann, London, 1910, pp. 17–22.

The worst journey in the world
Apsley Cherry-Garrard, *The Worst Journey in the World*: Carroll & Graf, London, 1922, pp. 234, 236–239, 265–272.

The hooligan cocks of Cape Adare
Douglas G.D. Russell, William J.L. Sladen and David G. Ainley, "Dr George Murray Levick (1876–1956): Unpublished notes on the sexual habits of the Adélie penguin", in *Polar Record*: Cambridge University Press, January 2012. Reproduced with the permission of the authors and the editor of *Polar Record*.

George Murray Levick and the Adélie Penguins
Helen Heath, *New Zealand Listener*, April 6, 2013, p. 41.

Impressions
Robert Falcon Scott, *Scott's Last Expedition: The Journals of Captain R.F. Scott*: John Murray, London, 1923, pp. 120–121.

Geologising on the Beardmore
Robert Falcon Scott, *Scott's Last Expedition: The Journals of Captain R.F. Scott*: John Murray, London, 1923, pp. 438–447.

Ice monsters, growlers and bergy bits
Raymond Priestley, "Ch XII: Antarctic Icebergs", in C.S. Wright and R.E. Priestley *Glaciology: British (Terra Nova) Antarctic Expedition 1910-1913*: Harrison & Sons, London, 1922, pp. 402–412.

Byrd makes a meteorological observation
Richard E. Byrd, *Alone: The Classic Polar Adventure*: Island Press, Washington, D.C., 2003, pp. 62–64, 146–156. Copyright 1938 by Richard E. Byrd, renewed 1966 Marie A. Byrd. Afterword copyright 2003 by Kieran Mulvaney. *Alone* was originally published by G.P. Putman's Sons, original text design by Paul Johnson.

Crabeaters and leopards
E.G. Turbott, "Chapter 7: Seals of the Southern Ocean" in *The Antarctic Today: A mid-century survey by the New Zealand Antarctic Society*, edited by Frank A. Simpson: A.H. and A.W. Reed, Auckland, 1952, pp. 203-207. Reprinted with the permission of the New Zealand Antarctic Society.

Innocents in the Dry Valleys
Colin Bull, *Innocents in the Dry Valleys*: Victoria University Press, Wellington, 2009, pp. 83-105. Review quote from *Quaternary Science Reviews* 29 (2010): pp. 3994–3996.

Birds and mammals of Antarctica
Bernard Stonehouse, "Chapter Six: Birds and Mammals" in *Antarctica* edited by Trevor Hatherton: A.H. and A.W. Reed, Auckland, 1965, pp. 153-156. Reprinted with the permission of the New Zealand Antarctic Society.

Catching falling stars
William A. Cassidy, *Meteorites, Ice, and Antarctica: A Personal Account*: Cambridge University Press, Cambridge, 2003, pp.1-3, 64–67, 124. Reprinted with the permission of Cambridge University Press.

The accidental penguin biologist
Lloyd Spencer Davis, *Professor Penguin: Discovery and Adventure with Penguins*: Random House, Auckland, 2014, pp. 37–40.

Sources and copyright

Food chain
Bill Manhire, *Collected Poems*: Victoria University Press, Wellington, 2001, pp. 271–272.

Krill
David G. Campbell, *The Crystal Desert: Summers in Antarctica*: Minerva, London, 1993, pp. 97–104.

An average day in the deep field
John Long, *Mountains of Madness*: Allen and Unwin, Sydney, 2000, pp. 92–95, 104–107.

Water, ice and stone
Bill Green, *Water, Ice & Stone: Science and Memory on the Antarctic Lakes*: Bellevue Literary Press, New York, 2008, pp. 106–111.

The Lakes of Mars
Chris Orsman, *The Lakes of Mars*: Auckland University Press, Auckland, 2008, pp. 35–37.

Katabatic winds
Stephanie Shipp, "Katabatic Winds" in *Curriculum Collections*: American Museum of Natural History: www.amnh.org/explore/curriculum-collections/antarctica/extreme-winds/katabatic-winds, 2002.

Inside the emperor penguin egg
Gavin Francis, *Empire Antarctica: Ice, Silence and Emperor Penguins*: Chatto & Windus, London, 2012 and Counterpoint, Berkeley, 2013, pp. 137–39. Reproduced by permission of The Random House Group Ltd and Counterpoint.

The mountains under the ice
Robin Bell, "Dispatches from the Bottom of the Earth: An Antarctic Expedition in Search of Lost Mountains Encased in Ice": *Scientific American*, November 12, 2008: http://www.scientificamerican.com/article/antarctic-expedition-in-search-of-lost-mountains/.

Antarctic time capsule
Michael S. Becker, "Diving an Antarctic Time Capsule": *The New York Times*, January 31, 2013: http://scientistatwork.blogs.nytimes.com/2013/01/31/diving-an-antarctic-time-capsule-filled-with-primordial-life/#more-22627.

Neutrinos on ice
Katie Mulrey, "Neutrinos on Ice", a survey of posts on *Scientific American*, 2014/2015: http://blogs.scientificamerican.com/expeditions/neutrinos-on-ice-detection-balloon-heads-to-antarctica/. Katie Mulrey was funded by the National Science Foundation.

Earth sans sunscreen
Jonathan Shanklin, "Unlayering of the Ozone: An Earth Sans Sunscreen": UN Chronicle, August 2009: http://unchronicle.un.org/article/unlayering-ozone-earth-sans-sunscreen/.

Waiting for the polar sunrise
Rhian Salmon, adapted from "An Average Day", originally published on felixsalmon.com, June 8, 2004, and now at http://smilingfootprints.com/2004/06/an-average-day/.

Life under ice
Paul Dayton and Simon Thrush, "Chapter 32: Antarctica" in *The Living Reef: The Ecology of New Zealand's Rocky Reefs*, edited by Neil Andrew and Malcolm Francis: Craig Potton Publishing, Nelson, 2003, pp. 256, 260–265.

Thirty-six million years in Antarctica
Rob Dunbar, adapted from a series of blog posts at *Ice Stories: Dispatches from Polar Scientists*: Exploratorium: http://icestories.exploratorium.edu/dispatches/category/antarctic-projects/wilkes-land-expedition/.

Hermaphrodite butterflies and acid seas
James McClintock, *Lost Antarctica: Adventures in a Disappearing Land*: Palgrave Macmillan, New York, 2012, pp. 117–138.

Sea ice and polynyas
Helen Bostock, adapted from a series on *Sciblogs: Field Work – Scientists on Expedition* during the Mertz Polynya voyage, with input from many of the voyage's participants: http://sciblogs.co.nz/field-work/2013/01/31/antarctic-voyage-the-mertz-polynya/, 2013.

The author would like to acknowledge the voyage leaders, Mike Williams and Steven Rintoul, the master, Evan Solly, and the experienced crew of the RV *Tangaroa*. The voyage was funded by scientific agencies in New Zealand, Australia and France.

What do seals want for Christmas?
Regina Eisert, adapted from "Update from the Ice": Antarctica New Zealand, November and December 2014: http://antarcticanz.govt.nz/images/downloads/science/141113UPDATE_FROM_THE_FIELD_Gone_Fish_Regina_Eisert.pdf

March of the king crabs
Kathryn Smith, adapted from "The march of the king crabs: a warning from Antarctica": *The Conversation*, July 2015: https://theconversation.com/the-march-of-the-king-crabs-a-warning-from-antarctica-43062.
Kathryn Smith was funded by the National Science Foundation.

Illustration credits

Front cover photograph: Mosaic satellite image of Antarctica, from NASA Earthdata, LANCE Rapid Response MODIS images and the NASA Level-1 & Atmosphere Archive and Distribution System (LAADS) Distributed Active Archive Center (DAAC), Goddard Space Flight Center, Greenbelt, MD.

The ice islands, January 9, 1773: William Hodges, Alexander Turnbull Library, Wellington, New Zealand, C-051-016   xxxii

Chinstrap penguins, c. 1820: Pavel Nikolaevich Mikhailov   12

Petrels, c. 1820: Pavel Nikolaevich Mikhailov   17

"Prise de possession de la Terre Adélie, le 21 Janvier 1840": Jules Dumont d'Urville, Macmillan Brown Library, Bib#1088742   20

The *Peacock* in contact with icebergs, 1840: Alfred Thomas Agate   26

HMS *Erebus* and *Terror* lying off Ross Island: Mary Evans Picture Library   36

South Polar Barrier, February 2, 1841: from *A Voyage of Discovery and Research in the Southern and Antarctic Regions during the Years 1839–43* by James Clark Ross, volume one: John Murray, London, 1847   40

Aurora australis, 1975–1976: Antarctica New Zealand Pictorial Collection   46

Members of British Antarctic Expedition man-hauling a sled, 1910–1913: Robert Falcon Scott   52

The *Discovery* ice-bound in front of Observation Hill, 1901: Herbert Ponting, Alexander Turnbull Library, Wellington, New Zealand, PAColl-6181-13   62

Borchgrevink's Hut, Cape Adare, August 1, 1902: C Beken collection, Canterbury Museum   73

Edward Wilson sketching on Beardmore Glacier, 1912: Robert Falcon Scott 76

Observation balloon next to *Gauss*, c. 1902: German South Polar Expedition 78

Sketch of secured balloon: Albert Stehr, from *The Southern Ice-Continent* by Erich von Drygalski, translated by M.M. Raraty: Bluntisham Books/ Erskine Press, 1989 80

Otto Nordenskjöld in expedition hut, Snow Hill Island, 1902: Otto Nordenskjöld Archive, Gothenburg University Library 84

Fern fossil *Cladophlebis* found at Hope Bay, c. 1904: Swedish Antarctic Expedition, from *Antarctica: Or Two Years Amongst the Ice of the South Pole* by Otto Nordenskjöld and Johan Gunnar Andersson: C Hurst & Company, London 1977 88

Toralf Grunden, Johan Gunnar Andersson and Samuel A. Duse at Snow Hill Island, 1903: Gösta Bodman 90

Alec L. Kennedy, a member of the Australasian Antarctic Expedition of 1911–14, using a Lloyd Creak dip circle, by Archibald Hoadley: Collections of State Library of New South Wales 96

Alistair Mackay, Edgeworth David and Douglas Mawson at the south magnetic pole, January 16, 1909: Edgeworth David 107

Photographs of a tardigrade and a nematode worm, 1910: from *British Antarctic Expedition 1907-9, Reports on the Scientific Investigations*, volume one, by James Murray: Heinemann, London, 1910 110

Drawings of a rotifer and a flagellatum 1910: from *British Antarctic Expedition 1907-9, Reports on the Scientific Investigations*, volume one, by James Murray: Heinemann, London, 1910 113

Emperor penguin eggs from Cape Crozier, c. 1911: Herbert Ponting 118

Sketch of emperor penguins at Barrier rookery, 1911: Edward Wilson 128

Edward Wilson, Henry Robertson Bowers and Apsley Cherry-Garrard eating a meal, August 1, 1911: Herbert Ponting, Alexander Turnbull Library, Wellington, New Zealand, PA1-f-067-043-2 131

Adélie penguins at Cape Adare, 1911–1912: George Murray Levick 134

George Murray Levick skinning a penguin, December 28, 1910: Herbert Ponting 138

Robert Falcon Scott writing in his journal, Cape Evans Hut, 1911: Herbert Ponting 142

Ponies pulling sleds on the Great Ice Barrier, December 2, 1911: Robert Falcon Scott, Alexander Turnbull Library, Wellington, New Zealand, PA1-f-066-05-02 144

Camp on Beardmore Glacier, 1912: Robert Falcon Scott, Alexander Turnbull Library, Wellington, New Zealand, PA1-f-066-03-7 146

Grotto in an iceberg, Plate CCLXXVII, 5 Jan, 1911: Herbert Ponting 158

The Matterhorn berg, Plate CCLXVI, 8 October, 1911: Herbert Ponting 163

Water-worn iceberg, Plate CCLXXXI, c. 1911: Hebert Ponting 171

Swan Ice on Cape Adare Ice-foot, Plate CCXVIII, c. 1911: Frank Debenham, Scott Polar Research Institute, Ref P54/16/616 172

Little America in the making, 1928: Richard Byrd 174

Richard Byrd using the telegraphy key at Advance Base, August 1934: Richard Byrd 180

Crabeater seal teeth: Craig Franklin 186

Colin Bull surveying from "Peak Alpha", 1958–1959: Colin Bull 196

Dick Barwick photographing a mummified seal, 1958–1959: Colin Bull 203

Pecten shell fossils 1958–1959: Colin Bull 206

Expedition campsite, Bull Pass 1958–1959: Colin Bull 208

Bernard Stonehouse with emperor penguins, Mt Riiser-Larsen ice shelf: Colin Monteath/Hedgehog House 216

Meteorite at Szabo Bluff, 2012: Dr Katherine Joy/Antarctic Search for Meteorites Program 222

Diagram of meteorite stranding surfaces: Field Museum 227

Lloyd Spencer Davis with Adélie penguin, Cape Bird: Lloyd Spencer Davis 230

Antarctic krill, *Euphausia superba*: Stephen Brookes/Australian Antarctic Division 238

Antarctic krill, *Euphausia superba*: Eric and Julie Hessell   243

John Long collecting fossils on Gorgon's Head, 1991: Margaret Bradshaw, Antarctica New Zealand Pictorial Collection   250

Fossils of a bothriolepis fish plate and lycopod stem, 1991–1992: Margaret Bradshaw, Antarctica New Zealand Pictorial Collection   256

Fraka Harmsen on Gorgon's Head, 1991: Margaret Bradshaw, Antarctica New Zealand Pictorial Collection   258

Lake Hoare campsite, McMurdo Dry Valleys, 1983: Bill Green   260

Lake Miers and Miers Glacier, 1975–1976: Terry Healy, Antarctica New Zealand Pictorial Collection   265

Whetter and Close fight through katabatic winds at Cape Denison, Australasian Antarctic Expedition, 1911–1914: Frank Hurley, National Library of Australia   272

Emperor penguin chicks: Gavin Francis   276

Twin Otter aircraft, East Antarctic Ice Sheet, 2008: Michael Studinger   280

Lake Untersee, January 28, 1996: Wilfried Bauer   288

ANITA III ready to be launched, November 2014: Ryan Nichol   292

ANITA III rolling out to launch site on "The Boss", January 5, 2015: Katie Mulrey   298

ANITA III balloon launching, January 5, 2015: Katie Mulrey   301

Ozone hole over Antarctica, October 3, 1985: NASA/Goddard Space Flight Center Scientific Visualization Studio   306

CASLab, Halley Station, 2004: Rhian Salmon/British Antarctic Survey   312

Rhian Salmon with icy eyelashes, 2004: Craig Nicholls/British Antarctic Survey   319

Rhian Salmon and Craig Nicholls up a mast, 2004: Alfonso Saiz Lopez/British Antarctic Survey   323

Sponges, anemones, and starfish in McMurdo Sound: Steve Alexander   328

Joides Resolution, November 12, 2012: Arito Sakaguchi/International Ocean Discovery Program JOIDES Resolution Science Operator (IODP JRSO)   334

Illustration credits

Sediment core from seabed off the coast of Wilkes Land, January 29, 2010: Rob Dunbar   341

Sea butterfly, *Limacina antarctica*: Alexander Semenov   346

Brittle star, Cape Bird, December 1971: Richard K McBride, Antarctica New Zealand Pictorial Collection   352

Sea urchin camouflaged with red seaweed, Cape Evans, 2001: Rod Budd, Antarctica New Zealand Pictorial Collection   353

Borch fish, *Pagothenia borchgrevinki*: Victoria Metcalf   356

Victoria Metcalf Jiffy drilling at Scott Base: Tessa Duder   361

Clarke Gayford ice fishing at Cape Evans: Victoria Metcalf   365

RV *Tangaroa* in Ross Sea, 2008: Peter Marriott/NIWA IPY-CAML   366

Nancy Bertler with ice core sample, Roosevelt Island, 2011: Nancy Bertler   374

Mobile fishing hut near Scott Base, 2014: Regina Eisert   380

Weddell seal eating a toothfish on the ice, December 2014: Regina Eisert   386

King crab on the Antarctic sea floor, February 2015: Allison Randolph   388

Sea lily and brittle stars on the continental shelf: Kathryn Smith   391

Acknowledgements

Firstly, my thanks go to Antarctica New Zealand, which enabled me to visit Antarctica in 2011 and 2014. From my 2011 trip, I would like to thank Matt Vance and Alice Miller, my companions on event K240. From my 2014 trip, I thank Cliff Atkins, my geologist colleague and team member on event K001-C, and the earth scientists on K001-B: Tim Naish, Richard Levy, Nick Golledge, Adam Lewis, Warren Dickinson, Andrew Gorman and Christoph Kraus.

Thanks also to all the Antarctic scientists and Antarctic Heritage Trust staff I met in New Zealand and on the ice for tolerating my questions and sharing their knowledge of Antarctica. And special thanks to the staff of Antarctica New Zealand in Christchurch and Scott Base for your warmth, generosity, wisdom and support.

Putting together an anthology like this takes not only a lot of research and reading time, but also a significant financial outlay to rights holders for both excerpts and images. I am indebted to the New Zealand Society of Authors, Copyright Licensing Limited and the Stout Research Centre for the 2010 Stout Centre Research Grant that gave me a three-month residency, with a stipend, to kick-start my ideas for this book. I would also like to thank the Ministry of Culture and Heritage for a 2013 New Zealand History Award that contributed both towards my time and to the payment of rights holders. For purchasing advance copies of the book and therefore helping fund the printing of this volume, my huge thanks to Antarctica New Zealand and the New Zealand Antarctic Research Centre at Victoria University of Wellington.

Thanks also to the patient team at Awa Press, particularly Mary Varnham and Emma Wolff. This book was a long time coming. Thanks, too, to Ruth Brassington, who assisted with word-for-word proofreads of my manuscript against original texts, and was my first outside reader and a valuable gauge of what a non-scientist accepted as a good read.

Being a writer and an academic demands time and energy away from my family. Thanks to Jonathan, Pippi, Hazel and Huck for tolerating my absences. Huck: I can't promise never to go to Antarctica again, but I do promise I will never visit the moon.

Finally, my warmest thanks to all the scientists, poets, photographers and other rights holders whose work is included in this anthology. Antarctica is the land of superlatives and a challenging place to write about. I'm inspired not just by your work on this frozen continent – often at temperatures much colder than the minus 20°C that floored me – but that you've each managed to capture something original and fresh in the way you present Antarctica.

R.P.

Index

Page numbers in **bold** refer to illustrations.

"An Account of the First Journey to the South Magnetic Pole" (David) 97–109, **107**
acidification, ocean 346, 347, 349–55, 363–64
Adams Glacier 262
Adamussium colbecki (Antarctic scallop) 331
Adélie Land **20**, 21–25, 223
Adélie penguins 68, 69, 71, 72, 74–75, 93, **134**, 239
  chicks 139
  Davies' account **230**, 231–35
  food requirements 381
  "George Murray Levick and the Adélie penguins" (Heath) 141
  sexual habits 135–40
Adineta grandis 112
Admiralty Bay 239–47
Advance Base hut, Byrd second expedition 175–85, **180**
Adventure (ship) 1, 2
Africa 87, 257
Agate, Alfred Thomas 26
Åkerlund, Gustaf 85
albatrosses 2, 16, 64, 65, 66
  black-browed *(Diomedea melanophyrs)* 64
  brown 7, 9, 10
  grey 3, 6
  yellow-nosed *(Thalassogeron culminatus)* 64, 65
algae 114, 116, 204, 217, 329, 333, 357, 363
alien species 389–91
Alone (Byrd) 174, 175–85, **180**
altimeter 209
altitude problems 100
ammonites 85, 88–89
amoebae 116
amphipods 244, 246
Amundsen Bay 216
Amundsen, Roald 47, 51, 147
anchor ice 332
Andersson, Johan Gunnar **90**, 91–95
Andrill marine sediment drilling project 375, 376
anemometer
  hand-held 197
  pole 175, 179, 181, 182, 183
ANITA (Antarctic Impulsive Transient Antenna) 293, 295
ANITA III **292**, 295–303, **298**, **301**
ANSMET (Antarctic Search for Meteorites) 222, 225
Antarctic (ship) 85, 90, 91, 92, 93, 95
Antarctic Circle 2, 13, 15, 47, 337, 338, 371

Antarctic Convergence 219, 240, 245
Antarctic Divergence 246
"Antarctic Icebergs" (Priestley) **158**, 159–73, **163**, **171**, **172**
Antarctic Manual for the Use of the Expedition of 1901, The (Murray (ed.)), 21–25, 53–61
Antarctic Penguins – A Study of Their Social Habits (Levick) 135
Antarctic Peninsula 13, 47, 84, 85, 91, 95, 351, 357, 392
Antarctic scallop *(Adamussium colbecki)* 331
Antarctic Today: A Mid-Century Survey by the New Zealand Antarctic Society, The (Simpson (ed.)), Chapter 7 **186**, 187–93
Antarctic Treaty 392
"Antarctica" (Dayton and Thrush) 329–33
Antarctica (Hatherton (ed.)), Chapter 6 217–21
Antarctica New Zealand 360, 377
Antarctica: Or Two Years Amongst the Ice of the South Pole (Nordenskjöld and Andersson) 85–89, 91–95
antifreeze glycoproteins 358
archaeocyaths 150
Archaeosigillaria 257
Archipel de Pointe-Géologie 25
Arctic summer sea ice 372
Arctowski, Henryk 48
Argentina 256
Armitage, Albert 76
Arnold, Tom 383–84
arthropods 114
  see also crustaceans; insects; mites
ascidians 329
Astrolabe (ship) 21–22, 24–25
atmospheric chemistry 320–24, **323**, 327
atmospheric pressure measurement 176
Aurora (ship) 367
aurora australis 31–32, **46**, 48–49, 50, 64
  Byrd's observations 175, 179–80
Aurora Australis (ship) 372
Australasian Antarctic Expedition, 1911–14 223, **272**
Australia 87, 191, 256, 336
"An Average Day" (Salmon) 313–27
Aztec Siltstone 252, 255, 258

bacteria 116, 213, 290–91, 349
Baker, Bill 354
balloon ascent by Drygalski **78**, 79–83, **80**
Barne, Michael 67, 76
barographs 176, 177, 181, 197
Bartley Glacier 199
Barwick, Dick 197, 200–01, 202, **203**, 204–14

Index

Bauguitte, Stéphane 312, 313
Bay of Whales 174, 175
Beardmore Glacier **52, 76,** 97, **146,** 147
Beaufort, Francis 38
Beaufort Island 38
Becker, Michael S. 288, 289–91
Belgian Antarctic Expedition, 1897–99 46–51
Belgica (ship) 47, 51, 191
Bell, Robin 281–87
Bellingshausen Sea 47
Bellingshausen, Thaddeus von 13–19
Bernacchi, Louis 72, 73, 76, 105, 106
bernach *(Trematomus bernacchii)* 331–32, 360, 362
Bertler, Nancy **374,** 375–79
Bertram, Colin 187, 188, 189, 191
Bird, Edward J. 41
"Birds and Mammals" (Stonehouse) 217–21
bivalves 331
blizzards 273
  Byrd's account 181–85
  Edgeworth David's account 100, 101, 105
  impact on seabirds 220
  John Long's field trip 253–55
  Scott's expedition 143, 144, 157
Blue Lake 114
Borchgrevink, Carsten 72, 73, 76
borch *(Pagothenia borchgrevinki)* 331–32, **356,** 359, 360, 362
Bostock, Helen 367–73
bothriolepis **256**
Bowers, Henry "Birdie" 118, 119–33, **131,** 147, 149
brachiopods, fossil 89
Bradshaw, Margaret 251, 253, 259
Bransfield Strait 245
British Antarctic (*Southern Cross*) Expedition, 1898–1900 72–74, **73**
British Antarctic (*Discovery*) Expedition, 1901–04 62, 63–77, 103, 105, 119, 128, 278
British Antarctic Expedition 1907–09, Reports on Scientific Investigations, vol. 1 111–16
British Antarctic Expedition, 1910–13 *see Terra Nova* Expedition, 1910–13
British Antarctic Survey 307, 308–09, 313, 320
  see also Halley Research Station
British Graham Land Expedition, 1934–37 191
brittle stars 351, 352, **352,** 353, 354, 355, 389, **391**
Brunt Ice Shelf 276, 307
bryozoans 329, 330
Bull, Colin **196,** 197–215
Bull, Gillian 212
Bull, Nicky 211
Bull Pass **206,** 206–07, **208,** 209–14
Byrd, Richard 174, 175–85, **180**

Cambrian period 169
cameras and photography 201, 204, 210, 212, 213, 237, 384, 385

Camp Ridley 162
Campbell, David G. 239–47
Campbell, Victor 139, 159
Canopus 31–32
Cape Adare 72–77, **73, 134,** 135–40, 158, 159, 162, 167, 172
Cape Bernacchi 162
Cape Bird 41, **230,** 231–35
Cape Crozier 40, 41, 118, 119, 120, 121, 125, 278
Cape Denison **272,** 273
Cape Evans 119, 121, 142, 159, **365**
Cape North 108
Cape Royds 97, 110, 111–16
carbon cycle 347–48
carbon dioxide levels 335, 343, 350, 351, 363–64, 375, 379
carbon dioxide sinks 371
carbonates 351, 352
Carboniferous period 256
Cassidy, William A. 223–29
Center for Remote Sensing of Ice Sheets, Lawrence, Kansas 284
cephalopods 188, 191, 192
CHABLIS (CHemistry of the Antarctic Boundary Layer and the Interface with Snow) 327
Challenger Expedition, 1872–76 161
Cherry-Garrard, Apsley 77, 118, 119–33, **131,** 277
China 225
Chionodraco hamatus (icefish) 357, 362, 364
chlorofluorocarbons (CFCs) 309, 310
Choetonotus 114
Christmas
  Cook's voyage 8
  Discovery Expedition 63–64, 68
  Scott Base 386, 387
ciliate infusiora 113
Clark, Bob 201
Clarke, Shaun 383–84, 385
Clear Lake 114
climate change
  history 85, 219, 318, 333, 336, 338–39, 340, 342, 343, 345, 371
  and ocean predators 384, 388–93
  present and future 310, 317–18, 333, 335, 336, 343, 352, 363–64, 372, 384, 388–93
  Roosevelt Island Climate Evolution (RICE) project **374,** 375–79
Clio helicina 28
Close, John **272**
clothing
  British Antarctic Expedition, 1910–13 56–57
  frozen 121–23
  furs 51, 57
  John Long's field trip 253
  Nimrod Expedition 101
  Rhian Salmon, Halley Research Station 314, 316, **319,** 325

Roosevelt Island Climate Evolution (RICE)
   project 377
Terra Nova Expedition 121–23
coal 150
Coast Lake 114
Cockburn Island 85
Columbia Scientific Balloon Facility 301
   Long Duration Balloon facility 296–303, **298**,
   301
comb jellies 330
Commonwealth Trans-Antarctic Expedition,
   1955–58 199
continental drift 157, 256–57, 336
Cook, Frederick 46, 47–51
Cook, James 159
   circumnavigation of Antarctica 1–11, 29
Cook Mountains **250**, 251
cooking 56, 91, 94–95, 129–30, 143, 210
copepods 244, 246, 357
corals 329, 351
cosmic microwave background (CMB) 294
cosmic ray detection 293, 294, 295, 296, 297,
   300, 303
crabs **388**, 389–90
Cretaceous period 86, 88
crevasses
   Cherry-Garrard's account 124, 125, 126, 131,
      132
   John Long's account 251
   Terra Nova expedition 144, 147, 148, 150,
      152, 153
Cross, Jacob 67
Crozier, Francis Rawdon Moira 37, 40–41
crustaceans 28, 30, 114, 189, 330
   Entomostraca 113
   fossils 88
   see also krill
The Crystal Desert: Summers in Antarctica
   (Campbell) 239–47
Cuce, Jacob 354–55
cuttlefish 191
cyanobacteria 290–91

Dais, Wright Valley 198, 200
Darby, John 232
Darwin Glacier 251
David, Edgeworth 97–109, **107**
Davis, Lloyd Spencer **230**, 231–35
Dayton, Paul 329–33
Deception Island 392
Deep Lake 114
Demidov, Dimitri 13, 14, 15
Denton Glacier 199
desmids 114
Detrich, Bill 354
Devonian period 251, 256, 257
DeVries, Art 358, 363, 381, 382, 383
*Diary of the Discovery Expedition to the
   Antarctic Regions, 1901–04* (Wilson) 63–77
diatoms 114, 330, 333

Difflugia vas 116
Diomedea melanophyrs (black-browed
   albatross) 64
Discovery (ship) 53, **62**, 63, 77
Discovery Expedition, 1901–04 **62**, 63–77, 103,
   105, 119, 128, 278
Discovery Hut, Ross Island 147
   see also Hut Point, Ross Island
Dispatches from the Bottom of the Earth
   (Bell) 281–87
"Diving an Antarctic Time Capsule Filled With
   Primordial Life" (Becker) 288, 289–91
diving under ice 289–91, 329–30
Dobson ozone spectrophometer 308
dogs, sledge-driving 60–61, 143, 199
dolerite 207, 213
Drake Passage 219
Dronning Maud Land 218
Dry Valleys *see* McMurdo Dry Valleys
Drygalski, Erich von 78, 79–83, **80**, 165
Dubouzet, Joseph 21–25
Dumont d'Urville, Jules 21–25, 73
Dumont d'Urville Station 25
Dunbar, Rob 335–45
Duse, Samuel **90**, 92, 94

East Antarctic Ice Sheet 281, 283, 284, 300, 335,
   375, 376
East Antarctica 223, 273, 277, 280, 281–87,
   311, 367, 377
echinoderms 217, 347–55, 389, 390
   see also individual species, e.g. starfish
eider ducks 59, 91
Eisert, Regina 380, 381–87
El Niño processes 332
emperor penguins 68–69, 70, 71, 114, 123, 124,
   126, 127–30, **128**, 301
   breeding 218, 221
   Brunt Ice Shelf **276**
   eggs **118**, 119–20, 127, 128–29, 130, 131,
      132–33, 277–79
   embryos 277–79
   Mt Riiser-Larsen Ice Shelf **216**
*Empire Antarctica: Ice, Silence and Emperor
   Penguins* (Francis) 277–79
Encrinus liliiformis 89
Enderby's Land 29
Endurance (ship) 191
Entomostraca 113
Erebus (ship) **36**, 37, 38, **40**, 41, 42
Erebus, Mount **36**, 38, 39, 41, 100, 199, 212,
   296
Euphausia 240, **243**, 246
   crystallorophias 240
   frigida 240
   glacialis 67
   superba 240–44
   triacantha 240
   vallentini 240
   see also krill

Index

Evans, Edgar 147–57
Evans, E.R.G.R. "Teddy" 149, 154
extremophiles 111, 289

Fabry, Victoria 349
Falkland Islands 193, 240
families *see* wives and families
Farman, Joseph 307, 309
fish 359, 363, 380, 381–83, **386**, 386–87
  fossils 250, 252, **256**, 257, 258–59
  frozen in icebergs 166–67
  McMurdo Sound 330, 331–32
  Notothenia 188, 357–60, 362–63, 364–65
  and sea butterflies 350
  seal diet 188, 189, 192, 363, 382, 383, **386**, 386–87
fishing
  commercial 381, 382, 392
  fieldwork 360, **361**, 362, 364, **380**, 381–84
  "Fishing in Antarctica" (Metcalf) 357–60, **361**, 362–65, **365**
flagellata 113, **113**, 115
Flying Fish (ship) 27, 34
food and food supply
  Belgian Antarctic Expedition 47
  Bellingshausen's voyage 13–14
  Bill Green's field trip 262
  British Antarctic Expedition, 1910–13 55–56
  Bull's expedition 197–98, 202, 204, 207, 209, 210, 211, 213, 214
  John Long's field trip 254, 255
  Nimrod Expedition 98, 99, 102, 104, 106, 108, 109
  Rhian Salmon, Halley Research Station 315, 325
  Swedish Antarctic Expedition 91–95
  Terra Nova Expedition 148, 149, 151–52, 153, 154, 155
  see also cooking
"Food Chain" (Manhire) 236–37
foraminifera 345
formaldehyde monitoring 321
Forster, Johann and Georg 1
fossils 85, 219, 252
  ammonites 85, 88–89
  archaeocyaths 150
  brachiopods 89
  Cladophlebis fern **88**
  crustaceans 88
  fish 250, 252, **256**, 257, 258–59
  Glossopteris fern 157
  mollusca 88, 89
  pecten shells **206**, 207
  plants 86, 87, 88, 89, 92, 255–57
  sea urchins 88
  vertebrates 86, 87, 219
Fox, John L. 32
France 20, 21, 22–23, 25
Francis, Gavin 277–79

frostnip and frostbite 58–59, 101, 120, 123, 147, 148, 156, 178, 179, 181, 317, 324, 326, 371
fulmars 3, 69
  slender-billed *(Priocella glacialoides)* 66, 69
Furneaux, Tobias 1

Gamburtsev Mountain expedition 281–87
Gardiner, Brian 307, 309
gas chromatography 321
Gastrotricha 114
Gauss 78, 79, 82, 83
Gaussberg 79, 81, 82, 83
Gayford, Clarke **365**
geology 23–24
  Belgian Antarctic Expedition 47
  Bull's expedition 197, 205, 211, 213
  Ross's expedition 38
  significance of icebergs 168–69
  Swedish Antarctic Expedition 85, 91
  Terra Nova Expedition 146, 149–50, 157
  Wilkes' expedition 27, 28
geophones 284
geophysics, subglacial lakes 281–87
Gerlache, Adrien de 47, 51
German South Polar Expedition, 1901–03 78, 79–83, 89
glacial–interglacial cycles 342–43
glacier icebergs 160, 161, 162, 164
Glacier Tongue 162
glaciers 24, 160, 169–70, 171, 214, **227**, 233, 331, 335, 372
  see also names of individual glaciers, e.g. Beardmore Glacier
Glaciology: British (Terra Nova) Antarctic Expedition 1910–1913 (Wright and Priestley) 158–73
global warming 310, 359–60
  see also climate change
Glossopteris fern 157
Gondwana 87, 157, 257
Goodspeed Glacier 199
Gorgons Head **250**, 251–52, 255, **258**, 258–59
Graham Land 188, 189
granite 146, 149, 207, **208**
gravity measurement 197, 212, 213, 281, 285
Great Ice Barrier *see* Ross Ice Shelf
Greenland Ice Sheet 375
Green, Bill 260–68
greenhouse gases 310, 343, 350, 351, 379
Grimminger, George 179
Grunden, Toralf **90**, 92, 94
gulls 59
Gunn, Bernie 199
Gymnodraco acuticeps (dragon fish) 362

Haines, William "Bill" 178–79
Halley Research Station 276, 277, 307, 308, 309, 313–27
  Clean Air Sector Laboratory **312**, 313, 316–17, 320–21, 324–26, 327

halons 309
Hamilton, J.E. 187, 192, 193
Hanson, Nicolai 73, 75, 187, 191
Haplostigma lineare 256
Harmsen, Fraka 251, 253, **258**, 259
Hart Glacier 199
Heath, Helen 141
health issues
  altitude problems 100
  blisters and peeling skin 59, 102–03, 120, 154
  Byrd 177–78, 179–80
  Evans' mental deterioration 155–56
  falls 147, 154, 156, 201
  snow blindness 59, 102, 154
  wounds and injuries 101, 147, 148
  see also frostnip and frostbite
The Heart of the Antarctic (Shackleton (ed.)),
  v. 2 97–109, **107**
helicopters 197, 212, 213, 231, 385
heliozoa 116
Homaxinella sponges 332
Hope Bay 88, **90**, 91–95
horsetail (lycopod) fossils 256–57
Hughes, Terence J. 197
humidity measurement 79, 176
Hut Point, Ross Island 119, 120, 384, 385
  see also Discovery Hut, Ross Island
Hutcheson, Guy 178
huts
  Advance Base hut, Byrd second
    expedition 175–85, **180**
  Cape Adare **73**, 73–74
  Discovery Hut, Ross Island 147
  Lloyd Spencer Davis 231–32
  Scott's Hut, Cape Evans **142**
  Swedish Antarctic Expedition **84**, **90**, 92, 95
hydrographical records 24, 197
hydroids 329
hygrometer 176

ice cores, analysis 317–20, 333, 376, 377–79
ice shelves
  collapses 364, 375
  Cook's expedition 29
  Wilkes' expedition **26**, 28, 29, 30–32, 33, 35
  see also Brunt Ice Shelf; Mt Riiser-Larsen Ice
    Shelf; Ross Ice Shelf
Ice Stories: Dispatches from Polar Scientists, blog
  posts by Rob Dunbar 335–45
ice thickness measurement 284–85
icebergs
  age of Antarctic bergs 164–65
  Arctic bergs 159, 160
  calved from Ross Ice Shelf 339, 369
  caves in bergs **158**, 166, 172
  coatings of sea spray 162–63, **163**, 166–67,
    171–72
  fish frozen in bergs 166–67
  geological significance 168–69

glacier bergs 160, 161, 162, 164
  Priestley's account **158**, 159–73, **163**, **171**, **172**
  Rob Dunbar's account 339–40
  "swan-ice" 170, **171**, **172**
  as water supply 16, 18–19, 28
  weathering and disintegrating features 166–71
icebergs seen by expedition members
  Bellingshausen's voyage 14, 15, 16, 18
  Challenger Expedition 161
  Cook's voyage 3, 4, 5–6, 7–8, 9, 10
  Discovery Expedition 66, 72, 77
  Dumont d'Urville's voyage 22, 24
  German South Polar Expedition 81–82
  Nimrod Expedition 161
  Ross's expedition 38, 161
  Terra Nova Expedition 161
  Wilkes' expedition 27, 28, 29–30, 31, 32,
    33–35
Wilkes Land Expedition 339
icefish *(Chionodraco hamatus)* 357, 362, 364
igloo building and shelter 124, 132
Ignatiev, Ivan 15
Infusoria 115
Innocents in the Dry Valleys (Bull) **196**, 197–215
insects 114, 217
Insel, Victoria Valley 198
Intergovernmental Panel on Climate Change
  (IPCC) 310
International Geophysical Year (IGY),
  1957–58 284, 308
International Polar Year (IPY) 281, 284
Intrepid (ship) 57
inversion winds 274
invertebrates 257, 390, 392
  see also individual invertebrates, e.g. mites
iodine oxide 324, 327

Jacquinot, Charles 21–25
Japanese glaciologists 223, 224–25
jellyfish 330, 350
Joides Resolution (ship) **334**, 335–45
Jonassen, Ole 85
"The Journal of M.J. Dumont-D'Urville" (extract
  by Dubouzet) 21–25
Jurassic period 88, 89, 92

katabatic winds **272**, 273–75, 368, 378
Kellett, Sir Henry 57
Kelly, Paul 387
Kemmerer water sampler 262, 264, 266, 268
king crabs **388**, 389–90
King George Island 239
Klövstad, Herlof 73
Koettlitz, Reginald 76
krill 30, 33, 67, 186, 190, **238**, 357, 362
  Campbell's account 239–47
  "Small fry" (Young) 248–49
  swarms 246–47
  see also Euphausia for species

Index

Lake Bonney 269–71
Lake Hoare 260
Lake Miers 261–65, **266**, 267–68
Lake Untersee **288**, 289–91
Lake Vanda 198, 201, 202
Lake Vashka 198
Lake Vida 198
Lake Vostok 283
"The lakes of Mars" (Orsman) 269–71
Larsen, Carl 85, 95
laser measurement of ice thickness 285
Lazarev, Mikhail 18
Lecointe, Georges 51
Leptolyngbya 291
Levick, George Murray 134, 135–37, **138**, 139–40
  "George Murray Levick and the Adélie penguins" (Heath) 141
lichens 204, 217
Limacina antarctica **346**, 347, 348
limestone 146, 149, 150, 169
limnology 260–68
Little America base, Byrd expeditions **174**, 175, 178
Living Reef: The Ecology of New Zealand's Rocky Reefs, The (Andrew and Francis (eds)), Chapter 32 329–33
Lloyd-Creak dip circle **96**, 97, 98, 103, 104, 105, 106, 108
Lobodon carcinophagus (crabeater seal) 67, 68, 69–70, 71, 72, **186**, 189–91, 192, 202, **203**, 218, 241
Long, John **250**, 251–59
Lost Antarctica: Adventures in a Disappearing Land (McClintock) 347–55
lycopod (horsetail) fossils 256–57
Lyeskov, Arkadiy 14, 18

Mackay, Alistair 97–109, **107**
Macquarie Island 193
Macrobiotus arcticus 113
magnetic observations 1, 14, 37, 38, 76, 97–109, 281, 285
  see also palaeomagnetic data
Malanzania 256
Manhire, Bill 143
  "Food Chain" 236–37
Marble Point airstrip 199
"The March of the King Crabs" (Smith) 389–93
marine birds *see* seabirds
marine mammals 63, 218, 219–21
  see also individual species, e.g. seals
Mario Zucchelli Station 357
Markham, Clements 53
Mawson, Douglas 97–109, **107**, 223, 272, 367–68
McClintock, James 347–55
McClintock, Sir F. Leopold 53–61
McKelvey, Barrie 197, 199, 200, 201, 205, 210–11, 212–13

McKelvey Valley 205, 207
McKenzie, Kate 383–84
McLoughlin, Steven 255–56
McMurdo Dry Valleys 196, 197–215
  lakes 260–64, **265**, 266–68
McMurdo Sound **62**, 63, 77, 207, **328**, 329–33, 357, 382, 383, 384, 385, 386
McMurdo Station 42, **292**, 296, 299, 349
"The Measure" (Miller) 117
Melville (ship) 246
Melville Island 57
Mertz Glacier 368, 369, 370
Mertz Polyna region 367, 369, 371, 372
Mertz, Xavier 368
Meserve Glacier 199
Metcalf, Victoria 357–60, **361**, 362–65
meteorites **222**, 223–29
  formation of stranding surfaces **227**, 227–28
Meteorites, Ice, and Antarctica (Cassidy) 223–29
meteorology
  Belgian Antarctic Expedition 47, 48
  Bull's expedition 197, 200, 209, 213
  Byrd's observations 175–85
microscopic animals and plants **110**, 111–16, 217, 290–91, 333
"Microscopic Life at Cape Royds" (Murray) 110, 111–16
Miers Glacier 262, **268**
Miers Valley 261–68
Mikhailov, Pavel 12, 15, 17
Miller, Alice 117
minimum thermometer 176, 177, 178, 179–80
Mirnyi (ship) 13, 14, 15, 18
mites 113, 114, 217, 218, 392
molluscs 217
  fossil 88, 89
Montreal Protocol 310
Monument Rock 156
Mormede, Sophie 381, 382, 383
mosses 112, 113, 116, 217, 392
Mount Bird 233
Mount Buckley 149, 150
Mount Darwin 147, 148, 149
Mount Erebus **36**, 38, 39, 41, 100, 199, 212, 296
Mount Huggins 212
Mount Jason 212
Mount Nansen 98
Mount New Zealand 100, 101
Mount Odin 200, 210
Mount Terror 38, 40, 41, 129
Mountain Mule rucksacks 204–05
Mountains of Madness (Long) 251–59
Mt Riiser-Larsen Ice Shelf **216**
Mulrey, Katie 293–303
Murphy, Charlie 178
Murray, James 110, 111–16

Narrative of the United States Exploring Expedition … (Wilkes), vol. 1 27–35
Naruse, Renji 224–25

Natural History Museum, London 118, 132–33
Natural History of the Adélie Penguin (Levick) 135
nematode worms 110, 111, 113, 114, 204, 217
neutrinos, high-energy 294–96, 297, 300, 303
"Neutrinos on Ice" (Mulrey) 293–303
New Harbour 331
New Year celebrations
  Discovery Expedition 65–66
  Nimrod Expedition 98–99
New York Times, The 288, 289–91
New Zealand 2, 119, 189, 191, 385
  see also Antarctica New Zealand; Scott Base
New Zealand Antarctic Division 231–32
New Zealand Antarctic Society 187, 387
Nicholls, Craig 323
Nichols, Bob 199
Nimrod Expedition, 1907–09 97–109, 107, 111, 159, 161
Ninnis, Belgrave 368
NIWA (National Institute of Water and Atmospheric Research) 367
Nordenskjöld, Otto 84, 85–89, 90, 92, 95
North Pole 51

Oates, Lawrence 147–57
"On Arctic Sledge-Travelling" (McClintock) 53–61
Orsman, Chris
  "The lakes of Mars" 269–71
  "The polar captain's wife" 43
ozone hole 306, 307–11, 372

pack ice
  Belgian Antarctic Expedition 47, 49, 51
  Cook's voyage 3–4, 6, 7, 10–11
  crabeater and Ross seals 189, 190–91
  Discovery Expedition 65, 66, 70, 71, 72, 77
  German South Polar Expedition 78, 79, 81
  icebergs 161–62, 164, 166, 170
  Lloyd Spencer Davis' account 233–34
  penguins 129
  seabird wintering 221
  seals 218
  Tangaroa in 366, 370, 371–72
"A page from the ice diary" (Bertler) 375–79
Pagothenia borchgrevinki (borch) 331–32, 356, 359, 360, 362
palaeomagnetic data 197, 213
palaeontology *see* fossils
Palmer Station 354
pancake ice 371–72
parasites 114
parhelion (mock sun) 102, 399
Parker, Steve 381, 382, 383
Parry Mountains 39
Parry, Sir William Edward 39
Paulet Island 95
Peacock (ship) 26, 27
"Peak Alpha" 196

Peary, Robert 51
pecten shell fossils 206, 207
pemmican 55, 60, 152, 209, 213
penguins 11, 16, 22, 23, 29–30, 50–51, 71, 72, 218, 219, 220
  catching 13–14, 93
  chinstrap *(Pygoscelis antarctica)* 12, 91, 239
  eggs 76, 86, 91 (*see also under* emperor penguins)
  as food 13–14, 47, 86, 91
  fossils 87, 219
  gentoo 239
  and krill 241, 247
  leopard seal predation 192, 235, 236–37
  recording of noises 79
  red-billed 2
  royal 47
  see also Adélie penguins; emperor penguins
Pensacola Mountains 223
peroxides measurement 321–22
Peter I Island 13
petrels 2, 3, 7, 16, 29, 220
  Antarctic *(Thalassoeca antarctica)* 3, 4, 6, 7, 8, 66, 68, 69
  blue 3, 8, 9, 17
  giant 68, 74, 239
  snow 67, 68, 69, 218
  white 17
  Wilson's 66, 68, 69, 75
  see also fulmars
petrified vegetation *see* fossils – plants
Philodina gregaria 112, 113
Phormidium 291
photography and cameras 201, 204, 210, 212, 213, 237, 384, 385
phytoplankton 247, 331, 371, 378, 379
pintadoes (cape pigeons/petrels) 2, 9, 29, 66
plate tectonics 336
Pleurogramma antarcticum (silverfish) 362, 363
Pliocene period 206, 207
polar amplification 343
"The polar captain's wife" (Orsman) 43
polar ice changes 342–43
Polar Plateau 53, 97, 147–48, 274
Polar Record 135–40
polar vortex 300, 309–10
polynyas 368, 370
ponies 143, 144
Porpoise (ship) 21, 27
porpoises 65, 66
Port Egmont hens 2
pressure ridges, ice 119, 120, 125–26, 151, 154, 156
Priestley, Raymond 158, 159–73
Priocella glacialoides (slender-billed fulmar) 66
Professor Penguin: Discovery and Adventure with Penguins (Davis) 231–35
proteridophytes 257
protozoa 217
psychrometers 79, 81

Index

ptarmigan 59
pteropod ooze 348
push ice 234
Pygoscelis antarctica (chinstrap penguin) **12**, 91, 239

Quadrula 116
quartz 150
quartz retro reflectors 321, **323**
Queen Elizabeth Land 223
Queen Fabiola (Yamato) Mountains 223–25
Queen Maud Land 223, 288, 289
Queen Maud Mountains **222**

radar, ice-penetrating 281, 284–85
radio communication
  ANITA III 295, 303
  Bill Green's field trip 261–62
  Bull's expedition 199–200, 205, 210–11, 213
  Byrd 175, 176, 178, 182
  John Long's field trip 254, 255
remote sensing technologies 281, 284–86
Resolute (ship) 57
Resolution (ship) 1, 2–11
rhizopods 111, 112, 113, 115–16
Robbers' Roost 233
Robertson Bay 77
rocks *see* geology; and specific rock types, e.g. granite
Ronne Antarctic Research Expedition, 1947–48 199
Roosevelt Island Climate Evolution (RICE) project 374, 375–79
Ross, Sir James Clark 37–42, 108, 159, 161
Ross Ice Shelf (Great Ice Barrier; Barrier) 53, 72, 160, 331, 364–65, 374
  Bay of Whales 174, 175
  Byrd's account 177, 179, 181, 182, 183
  climate change responses 375, 376, 379
  icebergs from 339, 369
  Long Duration Balloon facility 296, **298**, **301**
  Ross's account 39–40, **40**, 41
  Terra Nova Expedition 77, 119, 120–21, 125, 127, **128**, **144**, 147
Ross Island **36**, 37, 42, 63, 105, 119, 159, 329, 330
Ross Sea 71, 130, 161, 165, 331, 333, 357
  Antarctic bottom water formation 369
  fish 362, 363, 381
  pack ice **367**
  phytoplankton 378
  seals 188, 189, 384
Rossella racovitzae (hexactinellid sponge) 333
rotifers 111, 112, 114, 204, 217, 218
  bdelloid 112–13, **113**
Royal Geographical Society 53
Royal Society Range 212, 233, 234
Royds, Charles 66, 67, 69, 70, 71
rum 56

Russian Antarctic Expedition, 1819–21 12, 13–19

Sabine, Edward 103
Salmon Bay 331
Salmon, Rhian 312, 313–27, **319**, **323**
salpee 28
salps 246
sandstone 146, 149, 213
Sandwich Islands 15
sastrugi 54, 98, 99–100, 101, 103, 104, 105, 108, 148
satellite communication 296–97, 303
Sciblogs: Field Work – Scientists on Expedition Helen Bostock's blogs 367–73
Scotia Arc 219
Scott Base 42, 199, 205, 209, 231, 254, 255, 357, 360, **361**, 377, 381, 383, 384, 385
Scott, Robert Falcon 52, 62, 63, 73, 76, 119, **142**, 143–45, 147–57
Scott's Hut, Cape Evans **142**
Scott's Last Expedition: The Journals of Captain R.F. Scott 143–45, 147–57
scurvy 47, 51
sea anemones **328**, 329, 330
sea butterflies 330, 347–50
  Limacina antarctica **346**, 347, 348
sea elephants 30
sea ice formation and extent 233–34, 324, 338, 343, 358–59, 368–69, 370, 371–72, 379
sea leopards 48, 69
sea lilies 389, **391**
sea urchins 329, 351, **353**, 354
  fossil 88
seabirds 217, 218, 219, 220, 221, 350
  Discovery Expedition 63
  Forsters' discoveries 1
  leopard seal predation 192
  see also individual species, e.g. petrels
seals 2, 29, 47, 71, 86, 98, 217, 218, 219, 220, 221
  crabeater *(Lobodon carcinophagus)* 66–67, 68, 69–70, 71, 72, **186**, 189–91, 192, 202, **203**, 218, 241
  elephant 239
  fish as food 188, 189, 192, 363, 382, 383, **386**, 386–87
  as food 13–14, 47, 94
  leopard 192–93, 235, 236–37, 239
  mummified 197, 201, 202, **203**, 207, 211, 269
  Ross 69, 70–71, 191–92, 218
  "utlyuga" 15
  see also sea leopards; Weddell seals
"Seals of the Southern Ocean" (Turbott) **186**, 187–93
Seasonal Affective Disorder prevention 314
Secchi disk 262, 266–67
sediment cores, analysis 333, 334–45, **341**, 370–71

seismic data 209, 282, 284, 285, 286
Seward, Albert 157
The Sexual Habits of the Adélie Penguin
(Levick) 135
Seymour Island 85–89
Shackleton, Ernest 63, 64, 66, 68, 69, 70, 74,
107, 109
Shanklin, Jonathan 307–11
Shipp, Stephanie 273–75
shrimps *see* krill
Sideband, Mac 262
Silurian period 257
silverfish *(Pleurogramma antarcticum)* 362, 363
Simonov, Ivan M. 13, 14
Siple Dome 299
The Sisters 72
Skelton, Reginald 65, 66, 67, 70, 76
skidoos 251, 252, 253, 254, 255, 360
skiing and ski travel 68, 147, 148, 149, 151, 154,
155, 156
skuas 66, 75, 76, 98, 137, 204, 212, 233, 236,
239
McCormick's 68, 74, 114, 218
sledge travel
Bob Nichols 199
British Antarctic Expedition 1907–09 98,
99–102, 103, 104, 105, 108
Discovery Expedition 52, 53–61
John Long 251, 252–53, 254
Terra Nova Expedition 119, 120, 121, 123,
125–26, 132, 144, 147, 149, 150, 153, 154,
155, 156, 157
sleeping bags 49–51, 57, 121, 122, 143
"Small fry" (Young) 248–49
Smith, Kathryn 389–93
snow blindness 59, 102, 154
Snow Hill Island 84, 85, 88–89, 90, 92, 95
Souper, Constance 68
South America 87, 191, 256, 257
South Georgia 217
south magnetic pole 21, 27, 37, 38, 96, 97–109,
107
South Pole
first reached by Amundsen 51, 147
Scott's expedition 77, 119, 143, 147, 148, 149
Southern Cross (British Antarctic) Expedition,
1898–1900 72–74, 73
*The Southern Ice-Continent: The German South
Polar Expedition aboard the Gauss, 1901–03*
(Drygalski) 79–83
Southern Ocean
acidification 347–55
Antarctic bottom water 369, 370–71
Cook's observations 1
cyclones originating in 378
Discovery Expedition 63–72
fish 356, 357–60
German South Polar Expedition 78, 79
and glacial–interglacial cycles 343

icebergs 339–40
krill 244–45
monitoring between Wellington and Mertz
Polyna region 367, 368–73
sponges 217, 328, 329, 330, 332–33
glass 332
hexactinellid *(Rossella racovitzae)* 333
squid 188, 189, 191, 247
Staite, Brian 251, 253–54, 258
starfish 328, 329, 332, 351, 352, 353, 354, 355
Stehr, Albert 80
Stonehouse, Bernard 216, 217–21, 277
Strickland, Bill 387
stromatolites 291
Studinger, Michael 285
subantarctic islands 187, 188, 189
subglacial lakes 281–87
sun sights 200, 209
sunrise 323–24
Swedish Antarctic Expedition, 1901–04 84,
85–89, 90, 91–95
Syowa Base 224
Szabo Bluff 222

Tangaroa 366, 367, 370–72
tardigrades (water bears) 110, 111, 112, 113,
217
Tasmania 336
Taylor, Griffith 158
temperature inversion 274
temperature (air) measurement 78, 79, 81, 176,
177, 178, 335
tents 53, 101, 121, 122–23, 143, 205, 254–55,
260
terns 239
Terra Nova (ship) 158
Terra Nova Bay 158, 159, 331, 357, 362
Terra Nova Expedition, 1910–13
Beardmore Glacier and Scott's final march 76,
77, 146, 147–57
Cherry-Garrard's expedition to collect emperor
penguin eggs 118, 119–33, 131
Levick's account of Adélie penguins 134,
135–40, 141
McClintock's description of sledging 52, 53–61
Priestley's account of icebergs 158, 159–73,
163, 171, 172
Scott's impressions 142, 143–45
Terre Adélie 20, 21–25
Terror (ship) 36, 37, 40, 40
Tertiary period 86, 87, 88, 219
Thalassoeca antarctica (Antarctic petrel) 3, 4, 7,
8, 66, 68, 69
Thalassogeron culminatus (yellow-nosed
albatross) 64, 65
theodolite compass 97, 101, 103, 106, 108
theodolites 179, 200, 202, 204, 209, 210
thermographs 176, 177, 178–79, 197
thermometers 197
minimum 176, 177, 178, 179–80

Index

Thiel Mountains 223
Through the First Antarctic Night, 1898–1899
(Cook) 47–51
Thrush, Simon 329–33
toothfish 359, 363, 380, 381–83, **386**, 386–87
tourism 392, 393
Transantarctic Mountains 225, 251
Trematomus
    bernacchii (bernach) 331–32, 360, 362
    eulepidotus 362
    hansoni 362
    newnesi 362
    pennellii 362
Turbellaria 114
Turbott, Graham 186, 187–93
Turtle Rock 383–84, 385
Twin Otter aircraft **280**, 281

United Nations 310, 311
United States Exploring Expedition, 1838–42 **26**,
27–35
University of Canterbury 217, 232
    Gateway Antarctica 381
"Unlayering of the Ozone: An Earth Sans
Sunscreen" (Shanklin) 307–11
"Update from the Ice" (Eisert) 381–87

Varner, Larry 261, 262
ventifacts 207
Victoria Land 112, 196
Victoria University Antarctic expeditions 197–
214
Victoria Valley 198, 205
Vincennes (ship) 27
volcanic activity 38, 41, 42, 296, 378, 379
vorticellids 115
Vostok (ship) 12, 13–19
*The Voyage of Captain Bellingshausen to the
Antarctic Seas 1819–1821* 13–19
*A Voyage of Discovery and Research in the
Southern and Antarctic Regions during the
Years 1839–43* (Ross) 37–42
*A Voyage Towards the South Pole and Round the
World* (Cook) 1–11

Walker, J.D. 67
Warren, Guyon 199
water bears (tardigrades) **110**, 111, 112, 113,
217
*Water, Ice & Stone: Science and Memory on the
Antarctic Lakes* (Green) 261–68
water supply
    Bellingshausen's voyage 16, 18–19
    Cook's voyage 5
    Discovery Expedition 70
    John Long's field trip 254
    Nimrod Expedition 99
Webb, Peter 197, 198, 201, 205, 206, 207,
210–11, 212–13

Weddell, James 21
Weddell Sea 369
Weddell seals 72, 187–89, 239, 381, 382, 384
    counting and tagging 384–85
    toothfish as food 382, **386**, 386–87
Weller, William 67
Wenden, Max 231, 232
West Antarctic Ice Sheet 375, 376
whales 16, 29, 70, 192, 217, 218–19, 220
    baleen 241, 246, 247, 350
    blue 218, 239
    finback 31
    humpback 239
    hunting of whales 18
    killer 187, 191, 363, 381, 383
    right 30
    sperm 18
Whetter, Leslie **272**
Whillans Ice Stream 364–65
The Wide White Page: Writers Imagine Antarctica
(Manhire (ed.)) 143
Wiens, Doug 285–86
Wilkes, Charles 26, 27–35
Wilkes Land Expedition, 2010 334, 335–45, **341**
Wilson, Edward "Bill" 63–77, **76**, 118, 119–33,
131, 147–57, 187, 190, 277, 278
Wilson, Oriana (Ory) (née Souper) 63, 64, 65, 68
Wilson Piedmont Glacier 199
wind measurement and study 175–76, 177,
182–83, 273
    see also katabatic winds
wine 23
wintering over 93–95, 175–85, 313–27
wives and families
    Bull's wife and family 205, 211, 212
    Oriana Wilson (Ory) (née Souper) 63, 64, 65,
68
    "The polar captain's wife" 43
Woolfe, Ken 252, 258
worms
    Gastrotricha 114
    Nematoda **110**, 111, 113, 114, 204, 217
    Turbellaria 114
The Worst Journey in the World (Cherry-
Garrard) 77, 119–33
Wrangell, Ferdinand von 54
Wright, Charles **158**, 159
Wright Fiord 206, 207
Wright Lower Glacier 198, 199, 201
Wright Upper Glacier 198
Wright Valley 197–202, **203**, 204–05, 214

Yamato Mountains 223–25
Yeates, Peter 199–200, 209
Young, Ashleigh 248–49

Zélée 21–22, 24–25
zoophytes 28
zooplankton 244, 246